SpringerBriefs in Computer Science

More information about this series at http://www.springer.com/series/10028

Dev Oliver

Spatial Network Data

Concepts and Techniques for Summarization

 Springer

Dev Oliver
ESRI
Redlands, CA
USA

ISSN 2191-5768   ISSN 2191-5776   (electronic)
SpringerBriefs in Computer Science
ISBN 978-3-319-39620-0   ISBN 978-3-319-39621-7   (eBook)
DOI 10.1007/978-3-319-39621-7

Library of Congress Control Number: 2016941312

Printed on acid-free paper

This Springer imprint is published by Springer Nature
The registered company is Springer International Publishing AG Switzerland

To Datra and Drew

To Datra and Drew

Preface

Our daily lives often revolve around spatial networks such as transportation networks and utilities. Summarizing the activities that occur on these networks is of interest to professionals, organizations, and researchers in many domains including transportation safety, public safety, public health, and disaster response. For example, transportation planners and engineers may wish to identify road segments that pose risks for pedestrians and require redesign whereas law enforcement officials may desire to know which streets have increased crime activity in order to guide resource allocation decisions.

The process of summarizing spatial network data entails finding a compact description or representation of observations or activities on large spatial or spatiotemporal networks. However, summarizing spatial network data can be computationally challenging for various reasons, depending on the domain. This brief explores two of the main challenges: (1) the many connected components in the spatial network and (2) the many candidates that have to be processed. These challenges are conceptualized as well-defined problems and state-of-the-art techniques aimed at addressing the problems are discussed.

Redlands, CA, USA   Dev Oliver
April 2016

Acknowledgments

This brief would have not been possible without the support of Prof. Shashi Shekhar from the University of Minnesota. He supplied steady guidance, counsel, and feedback that helped to shape the overall direction of this brief.

Thanks to Springer for reviewing the manuscript. Thanks to Esri© for authoring the amazing software that produced some of the maps.

Special thanks to the following individuals for giving valuable insights into the societal applications of this work and for taking time to provide feedback on many topics discussed herein: Michael R. Evans, Xun Zou, Emre Eftelioglu, James M. Kang, Renee Laubscher, Veronica Carlan, Christopher Farah, Abdussalam Bannur, and Qiaodi Zhuang.

Lastly, I wish to express my deepest appreciation to my wife Datra, our son Drew, and our parents for their unwavering support during this journey.

Contents

Chapter 1
Introduction

Abstract This chapter presents the illustrative application domains for why spatial network data summarization is important as well as a summarization framework for different types or genres of data. The chapter ends by detailing the associated computational challenges.

Recent years have seen a number of human activities centered around spatial infrastructure networks, such as utilities (e.g., water, electricity), transportation, and oil/gas-pipelines. As a result, network-based location references such as street addresses are used in many activity reports including crime reports, pedestrian fatality reports, or reports on situational awareness. In many instances, the spatial interaction between activities at nearby locations are constrained by network distances (such as shortest paths along roads or train networks) and network connectivity rather than geometric distances (such as Euclidean or Manhattan distances used in traditional spatial analysis). Figure 1.1 shows an example of pedestrian fatalities occurring on a road network in Orlando, FL from 2010–2014 [1]. The pedestrian fatalities are not occurring arbitrarily in space but are instead constrained by network distance and network connectivity.

1.1 Summarizing Different Genres of Data

Although the characteristics of data may be different depending on its type or genre, the overall process of summarization tends to be similar across genres. Typically summarizing spatial network-based observations entails finding a compact description or representation of large spatial or spatio-temporal network datasets where the observations (e.g., pedestrian fatality reports, crime reports), are centered around the network. In general, the process involves defining a set of groups, finding a representative for each group, and reporting a statistic for each group (e.g., likelihood ratio, sum, mean). These ideas (group definition, representative, and statistic) differ depending on the genre or domain of the data being summarized.

© The Author(s) 2016
D. Oliver, *Spatial Network Data*, SpringerBriefs in Computer Science,
DOI 10.1007/978-3-319-39621-7_1

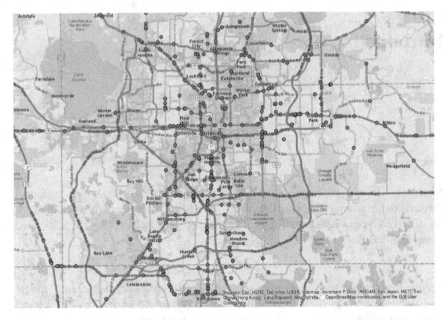

Fig. 1.1 Pedestrian fatalities occurring on arterials in Orlando, FL [1]. Activities such as pedestrian fatalities may be constrained by network connectivity and network distances (Best in color)

Table 1.1 Summarization framework for various data genres

Data Genre (Domain)	Group Definition	Group Representation	Interest Measure
Relational Table (a set of rows)	- A partition of rows - Significant groups of rows	- Attribute values (e.g., age-group)	- aggregate property - max coverage (e.g., count)
Vector Space	- A partition of vectors - Significant groups of vectors	- A vector, basis function	
Spatial (Euclidean Space)	- A partition of space - Significant groups of sub-space	- points, polygons, ellipses, line-strings	- significance metric (e.g., p-value, likelihood ratio, confidence interval)
Spatial Network	- A partition of a network - Significant groups of sub-networks	- path	- Computational metric - sequential response time
Geo-referenced Time-series (GT)	- A partition of a GT	- ST-full tree	

Table 1.1 outlines how different genres of data may be summarized. The genres included in the table are relational tables, vector space, spatial datasets, spatial networks, and geo-referenced time-series. The process of summarizing relational tables is akin to the GROUP BY clause in SQL, which is used to group rows having common values to report SQL aggregation functions such as mean and standard deviation [2]. The group definition is a partition of rows and the group representation may be distinct values of attributes such as income-group, citizenship, age-groups, etc. Similarly, the process of summarizing vector space involves defining the group, selecting a representative, and reporting a statistic; in this case the groups may be

a partition of vectors or significant groups of vectors, the group representative may be a vector or a basis function, and the statistic may be a significance metric (such as p-value, likelihood ratio, or confidence interval) or an aggregate property such as count, sum, or mean.

Summarizing spatial datasets (which often times use Euclidean distance) include techniques such as heat maps and hotspot analysis. Heat maps present a graphical representation of the data where individual values contained in a matrix are represented as colors. In the case of heat maps, the group definition might include a set of pixels and the group representation may be a subset of these pixels. Hotspots, on the other hand, are a special kind of partitioned pattern where objects in hotspot regions have high similarity in comparison to one another and are dissimilar to all the objects outside the hotspot [3]. The summaries in this case are based on spatial point locations where the group definition is a partition of space and the groups could be represented by points, polygons, ellipses, or line-strings. Figure 1.2a provides an example of a spatial summarization, where pedestrian fatalities in Orlando, FL are shown as dots and the spatial (Euclidean) summarization is represented using four ellipses. The spatial summarization technique used in this case was K-Means [4].

In the spatial network genre, the groups may be defined based on a partition of a graph and the group representative may be nodes, paths, trees, or subgraphs. An example may be seen in Fig. 1.2b. The thicker lines illustrate a summarization technique that uses linear representatives or paths to represent each partition [5]. Here four paths are used to summarize pedestrian fatalities (dots) in Orlando, FL that are on the spatial network (i.e., the road network).

The last genre of data listed in the table is geo-referenced time-series (GTs) [6]. Geo-referenced or geographic time-series allow us to observe the evolution in time of some phenomenon in a fixed location. Examples of GTs include the CIA World Factbook data for each year from 1989 to 2011 [7] and data on West Nile Virus cases for each of the 50 states in the US from 1999 to 2002. Summarization of geo-referenced time-series data may define groups based on a partition of a geo-referenced time series and may represent groups using spatio-temporal (ST)-nodes, ST-paths, or ST-heaviest full trees.

(a)   **(b)**

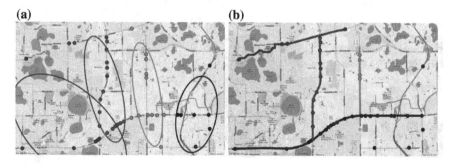

Fig. 1.2 An example of **a** spatial summarization (*ellipses*) and **b** spatial network summarization (*paths*) [5]

1.2 Illustrative Application Domains

Summarizing spatial network data is important in many domains including transportation safety, disaster response, environmental criminology, geo-politics, etc. Two applications that illustrate its societal importance are preventing fatalities and disaster response.

Preventing Pedestrian Fatalities: More than 47,700 pedestrians were killed in the United States from 2000 and 2009, according to a recent policy report [8]. During that same period, over 688,000 pedestrians were also injured, which equates to a pedestrian being struck by a vehicle every 7 min. Even though overall traffic deaths have fallen, Pedestrian fatalities have increased in many places, including 15 of the country's largest metro areas [8].

Pedestrian fatalities have largely been attributed to poor street design by domain experts; streets generally provide little or no provision for people on foot, riding bicycles, or in wheelchairs and have instead been engineered for speeding traffic [8]. Based on the emphasis on traffic movement, daily activities have shifted away from main streets towards higher speed arterials. As a result, more than half of fatal pedestrian crashes occur on these wide, high capacity and high-speed thoroughfares. Arterials are not built with pedestrians in mind and they are typically designed with four or more lanes and high travel speeds (Fig. 1.3a). They lack sidewalks, crosswalks (or have crosswalks spaced too far apart), pedestrian refuges, street lighting, and school and public bus shelters [8].

Serious consequences such as fatalities could arise due to this lack of basic infrastructure, e.g., forty percent of fatalities occurred where no crosswalk was available [8]. Figure 1.3b shows a map of pedestrian fatalities that occurred on Orlando roads from 2000–2009. Transportation planners and engineers need new spatial data summarization tools in order identify which frequently used road segments/stretches pose risks for pedestrians and consequently should be redesigned.

Disaster Response: In a disaster response scenario, immediate action is needed after a disastrous event to save lives, protect property, and deal with the immediate

(a)   **(b)**

Fig. 1.3 **a** Pedestrian at risk on a road without proper sidewalks [8]. **b** Pedestrian fatalities occurring on arterials in Orlando, FL [1]

Fig. 1.4 Emergency Requests (represented as *dots*) occurring in Port-au-Prince, Haiti after the 2010 Earthquake. (*Source* Crisis Map of Haiti [10])

disruption, damage or other effects caused by the disaster [9]. During the 2010 Haiti earthquake, disaster response was crucial in addressing the many requests for assistance in the form of food, water, and medical supplies [10]. Figure 1.4 shows emergency requests in Port-au-Prince, Haiti during this time [10]. Emergency managers need the means to summarize these requests so that they can better understand how to allocate relief supplies.

Table 1.2 shows a sample of the emergency request data [10]. The data includes the ID, title, date, location and category for each activity occurring during the earthquake. For example, activity 548 indicates that people were trapped at 1 Martin Luther King Avenue, Port-au-Prince. With the location, it is possible to plot and summarize emergency requests.

Table 1.2 Sample data showing activities in Haiti after the 2010 Earthquake [10]

Id	Title	Date	Location	Category
548	People trapped at GOC Universite	1/16/2010 5:21	1 Ave. Martin Luther King, Port-au-Prince	People trapped
1767	Fuel/Oil slick	1/23/2010 12:24	Port-au-Prince Harbor	Threats
1994	Need food/Water	1/23/2010 19:33	La Gonave	Water/Food shortage

1.3   Computational Challenges

Although useful and of great societal importance, summarizing the observations that occur on spatial networks is computationally challenging for the following reasons: (1) There may be a large number of k-subsets of connected components in the network (many connected components) and (2) There may be a large number of candidates (many candidates).

Many Connected Components: Finding a set of k connected components such as shortest paths is computationally challenging. This is due to the fact that if k connected components are selected from all connected components in a spatial network, there are a large number of possibilities for large k, i.e., $\binom{n}{k}$, where n is the number of connected components. This is because different subsets of k connected components could be overlapping or have the same connected components. For disjoint components, the problem would be relatively less computationally challenging. However, due to overlapping connected components, the general problem remains extremely challenging.

Many candidates: The search space itself may have potentially large number of candidates. For example, in a transportation planning scenario, there may be as many as 10^{16} shortest paths in a given dataset with hundreds of millions of activities or road network nodes. For large roadmaps such as the 100 million road-segments in the US, this results in prohibitive shortest path computation times.

This brief explores each of these challenges in the upcoming chapters.

References

1. Fatality Analysis Reporting System (FARS) Encyclopedia, National Highway Traffic Safety Administration (NHTSA). http://www.nhtsa.gov/FARS.
2. Elmasri, R. (2008). *Fundamentals of database systems*. Delhi: Pearson Education India.
3. Shekhar, S., Evans, M. R., Kang, J. M., & Mohan, P. (2011). Identifying patterns in spatial information: A survey of methods. *Wiley Interdisciplinary Reviews: Data Mining and Knowledge Discovery, 1*(3), 193–214.
4. MacQueen, J. (1967). Some methods for classification and analysis of multivariate observations. In *Proceedings of the fifth Berkeley symposium on mathematical statistics and probability* (Vol. 1 No. 14).
5. Oliver, D., Shekhar, S., Kang, J. M., Laubscher, R., Carlan, V., & Bannur, A. (2014). A k-main routes approach to spatial network activity summarization. *IEEE Transactions on Knowledge and Data Engineering, 26*(6), 1464–1478.
6. Oliver, D., Shekhar, S., Kang, J. M., Laubscher, R., Carlan, V., & Evans, M. R. (2012). Georeferenced time-series summarization using k-full trees: A summary of results. In *2012 IEEE 12th International Conference on Data Mining Workshops (ICDMW)*. IEEE.
7. CIA World Factbook. https://www.cia.gov/library/publications/the-world-factbook/
8. Ernst, M., Lang, M., & Davis, S. (2011). Dangerous by design: Solving the epidemic of preventable pedestrian deaths. Transportation for America: Surface Transportation Policy Partnership, Washington, DC.

9. Carter, W. N. (1991). *Disaster management: A disaster manager's handbook*. Manila: Asian Development Bank.
10. Crisis Map of Haiti. http://haiti.ushahidi.com/.

Chapter 2
Many Connected Components

Abstract This chapter presents different ways of handling the first challenge of summarizing spatial network data, i.e., the large number of k-subsets of connected components in the network. This challenge is conceptualized as the spatial network activity summarization problem (SNAS) where given a spatial network, a collection of activities and their locations (e.g., placed on a node or an edge), and a desired number of paths k, SNAS finds a set of k shortest paths that maximizes the sum of activities on the paths (counting activities that are on overlapping paths only once) and a partitioning of activities across the paths.

2.1 Introduction

Handling the large number of connected components in a spatial network may be conceptualized as the spatial network activity summarization problem (SNAS) [1]. This problem may be formalized in different ways depending on the domain or application. For the domain of summarizing pedestrian fatalities or crime analysis, one may define the problem as follows: Given a spatial network, a collection of activities and their locations (e.g., placed on a node or an edge), and a desired number of paths k, find a set of k shortest paths that maximizes the sum of activities on the paths (counting activities that are on overlapping paths only once) and a partitioning of activities across the paths. An activity may be the location of a carjacking, a pedestrian fatality, a train accident, or any other spatial network observation. In many domains or applications such as transportation planning, the aim is usually to help people get to their destination as fast as possible; SNAS, therefore, assumes that every path is a shortest path.

An example of the spatial network activity summarization (SNAS) problem is illustrated in Fig. 2.1. The input (Fig. 2.1a) consists of eight nodes, seven edges (with edge weights of 1 for simplicity), eleven activities, and $k = 2$, indicating that two routes and groups are desired. The shortest paths and the activity coverage (i.e., the sum of activities on each shortest path) for this network is shown in Table 2.1. The output (Fig. 2.1b) contains two shortest paths and two groups of activities. The shortest paths are representatives for each group and each shortest path maximizes

© The Author(s) 2016

D. Oliver, *Spatial Network Data*, SpringerBriefs in Computer Science,

DOI 10.1007/978-3-319-39621-7_2

Fig. 2.1 Example **a** Input
and **b** Output of Spatial
Network Activity
Summarization (Best in
color)

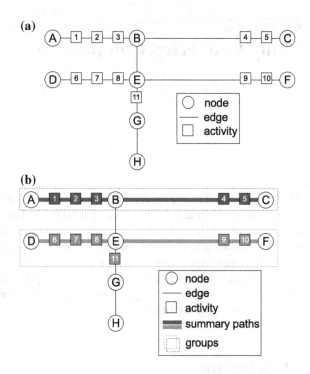

Table 2.1 Shortest paths from Fig. 2.1 (*Activity Coverage refers to the number of activities on a path*)

Source	Sink	Shortest path	Activity coverage	Source	Sink	Shortest path	Activity coverage
A	B	⟨A, B⟩	3	C	E	⟨C, B, E⟩	2
A	C	⟨A, B, C⟩	5	C	F	⟨C, B, E, F⟩	4
A	D	⟨A, B, E, D⟩	6	C	G	⟨C, B, E, G⟩	3
A	E	⟨A, B, E⟩	3	C	H	⟨C, B, E, G, H⟩	3
A	F	⟨A, B, E, F⟩	5	D	E	⟨D, E⟩	3
A	G	⟨A, B, E, G⟩	4	D	F	⟨D, E, F⟩	5
A	H	⟨A, B, E, G, H⟩	4	D	G	⟨D, E, G⟩	4
B	C	⟨B, C⟩	2	D	H	⟨D, E, G, H⟩	4
B	D	⟨B, E, D⟩	3	E	F	⟨E, F⟩	2
B	E	⟨B, E⟩	0	E	G	⟨E, G⟩	1
B	F	⟨B, E, F⟩	2	E	H	⟨E, G, H⟩	1
B	G	⟨B, E, G⟩	1	F	G	⟨F, E, G⟩	3
B	H	⟨B, E, G, H⟩	1	F	H	⟨F, E, G, H⟩	3
C	D	⟨C, B, E, D⟩	5	G	H	⟨G, H⟩	0

the activity coverage for the group it represents. For example, route $\langle A, B, C \rangle$ is the representative for the group comprised of activities 1, 2, 3, 4, and 5, and route $\langle D, E, F \rangle$ is the representative for the group comprised of activities 6, 7, 8, 9, 10, and 11.

SNAS may be applied in a variety of domains where observations happen along linear paths in the network. For example, crime analysts look for linear concentrations of crime (linear generators and attractors) such as speeding, street drug dealing, drunk driving, etc. to guide law enforcement [2]. Environmental engineers may try to summarize environmental change on water resources to understand the behavior of river networks and lakes [3]. Transportation officials are interested in understanding railroad accidents (e.g., derailing) to improve safety and reduce cost [4]. Emergency managers may find it useful to summarize the locations of stranded cars on highways to better understand how to allocate resources [5].

However, SNAS is computationally challenging due to the fact that if k shortest paths are selected from all shortest paths in a spatial network, there are a large number of possibilities for large k, i.e., $\binom{n}{k}$, where n is the number of shortest paths. The reason for this is that different subsets of k shortest paths could be overlapping or have the same shortest paths. If paths are disjoint, the computational challenge goes away but with overlapping paths, the general problem of SNAS is NP-complete (the proof is provided in Sect. 2.3).

2.1.1 An Illustrative Application Domain: Crime Analysis

As part of their analysis, crime analysts look for linear concentrations of crime such as speeding or drunk driving to help guide law enforcement decisions (e.g., deciding which streets to patrol). A key theoretical underpinning to crime analysis is Environmental Criminology, which is the study of crime, criminality, and victimization as they relate to particular places and to the way that individuals and organizations shape their activities spatially [6]. Environmental criminology is used by law enforcement to develop an understanding of the spatial distributions (e.g., high activity places or hot-spots) of crime activities as well as location-based factors affecting activities using analytical techniques, e.g., crime mapping, and analytical tools such as Crime-Stat [7]. Police departments also leverage knowledge of environmental criminology to facilitate the design of patrol routes and provide street-based descriptions of crime attractors and generators (e.g., a bar after closing time) [8]. Locations in crime reports reference symbolic systems (e.g., street addresses, highway-mile markers) as well as numerical systems (e.g., latitude/longitude), which refer to a point. Real crime report datasets collected by police departments in formats advocated by the US Department of Justice are illustrated in Table 2.2.

Spatial theories in Environmental Criminology include Routine Activity Theory (RAT) [9] and Crime Pattern Theory (CPT) [10]. RAT postulates that the location of a crime is related to the frequently visited areas of a criminal and CPT extends this theory on a spatial model. Crime is not spread evenly across maps but instead is

Table 2.2 Sample crime report data in formats advocated by the US Department of Justice

ID	Offence type	Date	Time	Address
96	Burglary	1/14/2006	1530	16950 GRAND AVE
477	Auto Theft	8/2/2006	2042	7950 W FAIRVIEW 1960
633	Narcotic Drug Laws	11/2/2006	1200	12950 CLEVELAND DR

Fig. 2.2 a Types of crime hot spots. **b** An example of a street-based (linear) hotspot of crime in a major US city

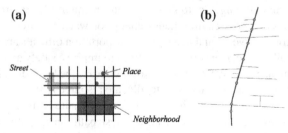

concentrated in some areas and absent in others [2]. This knowledge is used everyday by people and is seen in the way they avoid some places and seek out others. Police also use this understanding to make decisions about how to allocate scarce resources based in part on where police demand is highest. Officers are told to be attentive in certain areas but are not given guidance in other areas where crime is scarce [2].

Crime hotspots are not only studied in terms of places and neighborhoods, but also streets [2]. For places, an explanation as to why crime events occur at specific locations is sought. For neighborhoods, analysts look at large areas and ask questions such as "Which areas are claimed by gangs and which areas are not?". Street-based analysis deals with crimes that occur over small stretched areas such as streets or blocks and examples include street drug dealing, prostitution, and robberies of pedestrians [2]. Different types of crime hotspots are illustrated in Fig. 2.2.

The U.S. Department of Justice's research, development, and evaluation agency (i.e., the National Institute of Justice) points out that *"commonly available mapping programs make it easy to identify hot spot places or hot spot areas, but do not make linear hotspots easy to identify. Most clustering algorithms, unfortunately, will show areas of concentration even when a line is the most appropriate dimension"* [2].

2.1.2   State of the Art

Activity summarization is a significant area in data mining and spatial computing. Most techniques can be classified as being either geometry-based [11–15] or network-based [16–25].

Geometry-based summarization techniques involve grouping similar points distributed in planar space where distance is calculated using Euclidean distance, not network distance. Such techniques focus on the discovery of the geometry (e.g., circle, ellipse) of high density regions [2] and include K-Means [11], K-medoid [12, 13], P-median [14] and Nearest Neighbor Hierarchical Clustering [15]. These methods are useful for providing geometric summaries such as ellipses or circles but they may fail to group activities that occur on the same street.

Network-based summarization, on the other hand, involves grouping spatial objects using network (e.g., road) distance. Examples of network based summarization techniques include Mean Streets [16], Maximal Subgraph Finding (MSGF) [17], and Clumping [18–25]. These techniques may group activities over multiple paths, a single path/subgraph, or no paths at all. For example, Mean Streets [16] finds anomalous streets or routes with unusually high activity levels. It is not designed to summarize activities over k paths because the number of high crime streets returned is always relatively small. MSGF [17] identifies the maximal subgraph (e.g., a single path, $k = 1$) under the constraint of a user specified length and cannot summarize activities when $k > 1$. The Network-Based Variable-Distance Clumping Method (NT-VCM) [25] is an example of the clumping technique [18–25]. NT-VCM groups activities that are within a certain shortest path distance of each other on the network; therefore, a distance threshold must be specified.

Figure 2.3b illustrates an example output of NT-VCM. The threshold distance for NT-VCM is the unit distance between activities 8 and 11. NT-VCM is useful for summarizing activities on the network but the onus is placed on the user to specify a

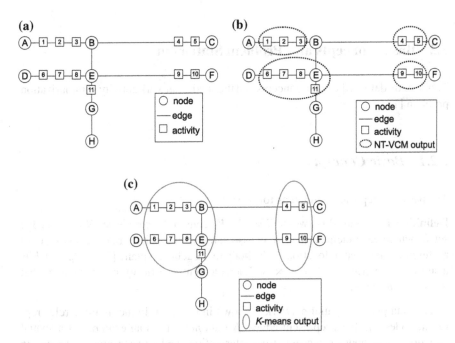

Fig. 2.3 Two methods of summarizing activities on a network. **a** Input. **b** NT-VCM Output [25]. **c** K-Means Output (network distance)

proper distance threshold. In this example, activities 1, 2, 3, 4, and 5, which occur on the same street would not belong to the same group, given this threshold distance.

Certain geometry-based summarization techniques may also be retrofitted to account for network distances. Figure 2.3c shows an example of the ellipses output by K-Means using network distance. The left ellipse groups activities 1, 2, 3, 6, 7, 8, and 11 whereas the right ellipse groups activities 4, 5, 9, and 10. By leveraging network distance, these techniques can be tailored to additional applications. However, even when generalized with network distances, these methods output point-based or ellipsoid-based groups, not paths or other network-based representatives such as nodes or subgraphs.

2.1.3   Outline of the Chapter

The rest of this chapter is organized as follows: Sect. 2.2 presents the basic concepts and problem statement of SNAS. The computational structure and a proof of why SNAS is NP-Complete is explained in Sect. 2.3. A recent approach to solving this problem, i.e., the K-Main Routes algorithm, is also presented. Section 2.4 presents a case study comparing geometry and network-based summarization techniques. A discussion is presented in Sects. 3.5 and 3.6 summarizes the chapter.

2.2   Basic Concepts and Problem Statement

This section defines the basic concepts and the spatial network activity summarization problem [1].

2.2.1   Basic Concepts

The basic concepts are defined as follows:

Definition 1 A **spatial network** $G = (N, E)$ consists of a node set N and an edge set E, where each element u in N is associated with a pair of real numbers (x, y) representing the spatial location of the node in a Euclidean plane [26]. Edge set E is a subset of the cross product $N \times N$. Each element $e = (u, v)$ in E is an edge that joins node u to node v.

An example of a spatial network is shown in Fig. 2.1a. In the figure, circles represent nodes and lines represent edges. A road network is an example of a spatial network where nodes represent street intersections and edges represent streets. In utility networks such as electric networks, nodes represent devices such as transformers and fuses whereas edges represent lines (e.g., medium voltage lines) [27].

Definition 2 An **activity set** A is a collection of activities. An **activity** $a \in A$ is an object of interest associated with only one edge $e \in E$ or one node $n \in N$.

In Fig. 2.1a, activities are represented as squares. In transportation planning, an activity may be the location of a pedestrian fatality; in crime analysis, an activity may be the location of a theft; and in disaster response an activity may be the location of a request for relief supplies.

Definition 3 A **summary path set** $\hat{P}$ is a collection of summary paths where each path $p_i \in \hat{P}$ is a shortest path. A **summary path** imposes a partitioning on an activity set A such that $network\, distance(a, p_i) \leq network\, distance(a, p_j) \forall p_j \in \hat{P}, \forall a \in A$.

Figure 2.1b shows two summary paths $\langle A, B, C \rangle$ and $\langle D, E, F \rangle$. Activities 1, 2, 3, 4, and 5 form a partition around $\langle A, B, C \rangle$ because they are closer to $\langle A, B, C \rangle$ whereas activities 6, 7, 8, 9, 10 and 11 form a partition around $\langle D, E, F \rangle$ because they are closer to $\langle D, E, F \rangle$. Here the network distance between an activity and a path, i.e., $network\, distance(a, p_i)$, is the network distance between a and the closest node in p_i.

Definition 4 The **activity coverage** $AC(p)$ **of a path** p is the sum of activities having network distance $= 0$ from an edge $e \in p$. The **activity coverage** $AC(P)$ **of a set of paths** P is the sum of activities across individual paths p_i in set P, having network distance $= 0$ from each edge $e \in P$, counting activities that are covered several times only once. If two paths share an edge, the activities on that edge are only counted once.

An example may be seen in in Fig. 2.1a. Here the activity coverage of the shortest path from node A to node B is 3 because there are 3 activities occurring on that path. The activity coverage of the shortest path from node D to node F is 5 because there are 5 activities occurring on $\langle D, E, F \rangle$. If the set of paths are $\langle A, B, C \rangle$ and $\langle D, E, F \rangle$, then the activity coverage is 10 because there are 10 activities on all the edges of the paths in P. If the set of paths are $\langle A, B, C \rangle$ and $\langle A, B \rangle$, then the activity coverage is 5. The activities on edge AB are only counted once even though AB is an edge in both paths.

2.2.2 Problem Statement

The problem of spatial network activity summarization (SNAS) can be expressed as follows:

Given:

1. A spatial network $G = (N, E)$ with weight function $w(u, v) \geq 0$ for each edge $e = (u, v) \in E$ (e.g., network distance),

2. A set of activities A and their locations (e.g., a node or an edge),
3. A desired number of summary paths, k, where $k \geq 1$.

Find:

1. A summary path set of size k,
2. A partitioning of activities across these summary paths.

Objective: Maximize the activity coverage of each summary path for the group it represents.

Constraints:

1. Each summary path is a shortest path between its end-nodes,
2. Each activity $a \in A$ is associated with only one edge $e \in E$.

In order to understand the problem statement, consider the inputs and the outputs to the problem. For the inputs, Definition 1 defines the spatial network, the activities represent objects of interest such as the locations of crimes, and k represents the desired number of summary paths. The outputs for SNAS are a summary path set of size k and a partitioning of activities across the paths. The summary paths are representatives for each group and each summary path maximizes the activity coverage for the group it represents.

Example. Figure 2.1a shows an example of the input, i.e., a spatial network such as a road network, composed of streets (edges) and intersections (nodes) with eleven activities (squares) and a specified value of $k = 2$. The goal is find two groups of activities and two routes, each route being the representative for each group. In a crime analysis scenario, identifying such routes would guide patrol efforts to mitigate crime on certain streets whereas in a transportation planning scenario, identifying such routes would guide street redesign efforts to reduce the risk of pedestrian fatalities (e.g., adding sidewalks, crosswalks, pedestrian refuges, street lighting, etc.). Figure 2.1b shows an example of the output. In this case, route $\langle A, B, C \rangle$ is the representative for the group comprised of activities 1, 2, 3, 4, and 5; and route $\langle D, E, F \rangle$ is the representative for the group comprised of activities 6, 7, 8, 9, 10 and 11.

When handling the challenge of many connected components, the domain often drives the way the problem is formulated. For the domains of crime analysis and transportation planning, the problem statement in its current form may be suitable to address the needs of law enforcement professionals and transportation planners. This is because maximizing activity coverage works well in these domains (e.g., finding the representative routes that cover most crimes in a group). However, for the domain of disaster response, maximizing activity coverage may not be the only desirable objective. An alternative objective in this case might be to minimize the distance that victims have to walk in order to receive assistance in the form of water, food, or medicine. The new objective may have an impact on the computational structure of the problem and different techniques may have to be employed to bring about a computationally efficient and correct solution. In the next section, the computational structure of SNAS is outlined.

2.3 Spatial Network Activity Summarization

This section describes the computational structure of SNAS. It also describes a new trend in this area, i.e., the K-Main Routes (KMR) algorithm and its performance-tuning decisions Network Voronoi activity Assignment, Divide and conquer Summary PAth REcomputation, and Inactive Node Pruning.

2.3.1 Computational Structure of Spatial Network Activity Summarization

In SNAS, the optimal solution may not be unique. Additionally, among the optimal solutions there are some where every path starts and ends at active nodes. These properties are formally shown via Lemmas 2.1 and 2.2.

Definition 5 An **active edge** is an edge $e \in E$ that has 1 or more activities. An **active node** is a node u joined by an active edge or a node that has one or more activities, or both. An **inactive node** is a node that is not joined by any active edges.

Edges AB and BC in Fig. 2.1a are active edges because they each have at least one activity and nodes A, B, C, D, E, F, and G are all active nodes because they are all joined by active edges. By contrast, Node H is an inactive node because it is not joined by any active edges.

Lemma 2.1 *The optimal solution for SNAS may not be unique.*

Proof There may be multiple solutions for different values of k. For example, given $k = 1$ in Fig. 2.1a where all eleven activities are members of the same group, the summary path could be $\langle A, B, E, D \rangle$ or $\langle D, E, B, A \rangle$, since both these paths have a maximum activity coverage of 6 based on the one group. Given $k = 2$ and the groups shown in Fig. 2.1b, the summary paths could be $\langle A, B, C \rangle$ and $\langle D, E, F \rangle$ or $\langle C, B, A \rangle$ and $\langle F, E, D \rangle$ as both sets of paths have a maximum activity coverage of 11 based on their respective groups.

Lemma 2.2 *Among the optimal solutions for SNAS, there exists optimal solutions where every path starts and ends at active nodes.*

Proof Let's begin with an arbitrary optimal solution. Let p be a shortest path that starts or ends with inactive nodes. If inactive nodes that start or end p are removed such that p starts and ends with active nodes, the resulting subpath p' is still optimal in terms of activity coverage, because no active edges were removed. p' is also still a shortest path due to the optimal substructure of shortest paths wherein subpaths of shortest paths are shortest paths [28]. In other words, eliminating inactive nodes from the beginning and end of a shortest path does not reduce coverage and does not split the path. KMR takes advantage of this property to achieve computational savings.

2.3.2 Proof of NP-Completeness

For simplicity, a generalized decision version of SNAS where the set of paths might be
arbitrary is first defined and then shown to be NP-complete. A proof sketch showing
that the decision version of SNAS with shortest paths is also NP-complete is then
presented.

Definition 6 Decision version of SNAS:

   INSTANCE: A spatial network $G = (N, E)$ with weight function $w(u, v) \geq 0$
for each edge $e = (u, v) \in E$, a set of activities A and their locations, a set of paths
P, a desired number of routes k, and a bound $B \in Z^+$, where Z^+ denotes the set of
positive integers.

   QUESTION: Does P contain a cardinality k subset P' of P, i.e., a subset $P' \subseteq
P$ with $|P'| = k$ and $AC(P') \geq B$, where $AC(P')$ denotes the activity coverage of
P'?

Theorem 2.1 *SNAS is NP-complete.*

Proof The process of devising an NP-completeness proof for a decision problem Π
consists of the following four steps [29]:

1. showing that Π is in NP,
2. selecting a known NP-complete problem Π',
3. constructing a transformation f from Π to Π', and
4. proving that f is a (polynomial) transformation

   In step 1, to show that SNAS $\in$ NP, assume that a certificate and a number B are
given. The certificate consists of a spatial network $G = (N, E)$ with weight function
$w(u, v) \geq 0$ for each edge $e_i = (u, v) \in E$, a set of activities A, a set of paths P, a
desired number of routes k, and a cardinality k subset P' of P, i.e., a subset $P' \subseteq P$
with $|P'| = k$. We can then verify in polynomial time whether $AC(P') \geq B$ because
$AC(P')$ involves counting the number of activities in P'.

   Step 2 selects the Maximum Coverage problem [30] as a known NP-complete
problem Π'. Although the optimization version of Maximum Coverage [30] is known
to be NP-Hard, its decision version is NP-Complete. The decision version is specified
as follows:

   INSTANCE: A number k and a collection of sets $S = S_1, S_2, \ldots, S_m$, where
$S_i \subseteq \{l_1, l_2, \ldots, l_n\}$, and a bound $B \in Z^+$, where Z^+ denotes the positive integers.

   QUESTION: Does S contain a subset $S' \subseteq S$ of sets such that $|S'| \leq k$ and the
number of covered elements $\left| \bigcup_{S_i \in S'} S_i \right| \geq B$?

   Steps 3 and 4 construct a transformation f from Π to Π' and prove that f is a
(polynomial) transformation. The reduction entails a polynomial time transformation
of the input of Maximum Coverage to the input of SNAS followed by a polynomial
time transformation of the output of SNAS to the output of Maximum Coverage.
The input of Maximum Coverage may be transformed to the input of SNAS using
the following steps:

1. Impose a total order TO on n elements in $L = \{l_1, l_2, ..., l_n\}$.
2. Convert each element in L into a node with one activity.
3. Convert each set S_i to a path P_i .

   - Add edge $(l_j, l_{j+1})\ \forall\ j \in 1...\, |S_i|$.

The transformation computation time is dominated by the polynomial step of sorting the elements in S_i using TO. Thus it is easy to see that the entire transformation is indeed polynomial. Consider an instance of the Maximum Coverage problem as shown in Fig. 2.4a where $L = \{l_1, l_2, l_3, l_5, l_6\}$, $k = 2$, $S_1 = \{l_1, l_2\}$, $S_2 = \{l_2, l_3\}$, $S_3 = \{l_1, l_2, l_3\}$, and $S_4 = \{l_5, l_6\}$. The resulting instance of SNAS would thus be $P = \{(l_1 \rightarrow l_2), (l_2 \rightarrow l_3), (l_1 \rightarrow l_2 \rightarrow l_3), (l_5 \rightarrow l_6)\}$, $k = 2$, $A = \{a_1, a_2, a_3, a_5, a_6\}$, and $ActivityNode = \{a_1(l_1), a_2(l_2), a_3(l_3), a_5(l_5), a_6(l_6)\}$.

Next, we convert the instance of SNAS output to an instance of Maximum Coverage output. The transformation, which is also polynomial, is as follows:

- For each k path P_i produced by SNAS, convert the activities on the path into elements and form a set S_i.

The candidate solutions for the instance of SNAS shown in Fig. 2.4a are $(l_1 \rightarrow l_2 \rightarrow l_3)$ and $(l_5 \rightarrow l_6)$. The resulting instance of the Maximum Coverage output would be $S_1 = \{l_1, l_2, l_3\}$ and $S_2 = \{l_5, l_6\}$.

The reduction of Maximum Coverage to SNAS is a polynomial time reduction since the input of Maximum Coverage can be reduced to SNAS in polynomial time, and the output of SNAS can be reduced to Maximum Coverage in polynomial time.

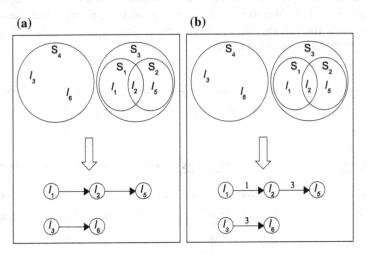

Fig. 2.4 SNAS instance resulting from Maximum Coverage instance for **a** arbitrary paths and **b** shortest paths

Since SNAS belongs to the class of NP and a known NP-complete problem is reduced to it, the decision version of SNAS is NP-complete.

We now present a proof sketch that the decision version of SNAS where the set of paths are shortest paths is also NP-complete. We follow the same construction as before but assign a cost for the edge (l_i, l_j) to be $j - i$ (Fig. 2.4b). This step ensures that all paths generated by construction are shortest paths.

2.3.3   Trend: The K-Main Routes Algorithm

The pseudocode for the K-Main Routes (KMR) algorithm that addresses SNAS is shown in Algorithm 1. KMR's basic structure resembles that of K-Means [11] in terms of selecting initial seeds, forming k groups, and updating group representatives until the assignments no longer change. In Line 1 of the algorithm, all shortest paths P that start and end with active nodes (inactive node pruning) are selected. In Line 2, k paths from P are selected as initial summary paths, which are the "seeds" for the algorithm. KMR then proceeds in two main phases. In the first phase, k groups are formed by assigning each activity to its closest summary path (Line 4). In the second phase, the summary path of each group is updated by calculating the shortest path that maximizes activity coverage (Line 5). These phases of assigning and updating repeat until the summary paths no longer change and the final summary paths and groups are returned (Line 8).

KMR Example: An example execution of KMR from first principles is shown in Fig. 2.5. In this example, the spatial network shown has eight nodes, seven edges, and eleven activities. For illustration purposes, $\langle A, B \rangle$ and $\langle D, E \rangle$ are selected as initial summary paths (though any set of paths may be used as the initial seeds), $k = 2$, and all edge weights are set to 1 (Fig. 2.5).

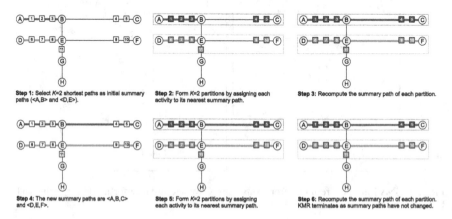

Fig. 2.5 Execution trace of K-Main Routes (KMR). *Circles* represent nodes, *lines* represent edges, and *squares* represent activities (Best in color)

Two groups are formed by assigning each activity to its closest summary path in step 2 of Fig. 3.3. Activities 1, 2, 3, 4, and 5 are assigned to summary path $\langle A, B \rangle$, and activities 6, 7, 8, 9, 10 and 11 are assigned to summary path $\langle D, E \rangle$. Dashed lines are used to highlight the groups. If an activity's distance to multiple summary paths is equal, the activity is randomly assigned to one of the paths.

The next step involves recalculating the summary paths of each group once the groups have been formed (step 3). In the example, the new summary paths (shown in step 4) are $\langle A, B, C \rangle$ and $\langle D, E, F \rangle$. As can be seen, these summary paths further maximize activity coverage, and this is the reason they are chosen.

In Step 5, the process of forming groups is repeated, but this time, the summary paths used are not the initial seeds we started with but the new summary paths $\langle A, B, C \rangle$ and $\langle D, E, F \rangle$ that we calculated in step 4. Again, dashed lines are used to highlight the groups. Another recalculation of the group representatives or summary paths is done in Step 6 to see if we can further maximize activity coverage. At this point, the algorithm determines that the summary paths that are recalculated do not change and as a result the algorithm terminates. We now take a closer look at each phase of the algorithm.

Algorithm 1 K-Main Routes (KMR) Algorithm

Input:
    1) a spatial network $G = (N, E)$,
    2) a set of activities A,
    3) a number of routes k,
    4) model $\in \{naive, NOVA\}$,
    5) mode2 $\in \{naive, D\text{-}SPARE\}$

Output:
    A summary path set of size k and a partitioning of activities across these summary paths, where the objective is to maximize the activity coverage of each summary path for the group it represents.

Algorithm:
1: $P \leftarrow$ shortest paths between active nodes of G
2: $\hat{P} \leftarrow k$ summary paths $\in P$; $stableGroups \leftarrow false$;
3: **while** not $stableGroups$ **do**
4:     **Phase 1**: $currentGroups \leftarrow AssignActivities\text{-}$
      $ToSummaryPaths(G, A, k, \hat{P}, model)$
5:     **Phase 2**: $\hat{P}' \leftarrow RecomputeSummaryPaths$
      $(G, A, k, currentGroups, mode2)$
6:     **if** $\hat{P} = \hat{P}'$ **then** $stableGroups \leftarrow true$
7:     **end if**
8:     $\hat{P} \leftarrow \hat{P}'$
9: **end while**
10: **return** $currentGroups$

2.3.3.1 Phase 1: Assign Activities to Nearest Summary Paths (i.e., Forming Groups)

This phase involves forming k groups by assigning each activity to its closest summary path. The pseudocode for the activity assignment algorithm is presented in Algorithm 2. The algorithm has two modes: naive and NOVA (Network Voronoi activity Assignment). In the naive mode, all distances between each activity and summary path are enumerated in order to determine the closes summary path. In contrast, NOVA avoids this enumeration and still delivers correct results.

Performance-Tuning for Phase 1: The Network Voronoi activity Assignment (NOVA) technique is a faster way of grouping activities, i.e., assigning each activity to its closest summary path. To understand NOVA, imagine a virtual node V was connected to every node in all summary paths by edges of weight zero (see Fig. 2.6). NOVA calculates the shortest path from V to all active nodes and discovers the closest

---

Algorithm 2 AssignActivitiesToSummaryPaths

Input:
    1) a spatial network $G = (N, E)$,
    2) a set of activities A,
    3) a number of routes k,
    4) a set of summary paths $\hat{P}$,
    5) mode $\in \{naive, NOVA\}$
Output:
    k groups formed by assigning each activity $\in A$ to the closest summary path $\in \hat{P}$.
Algorithm:
1: **if** mode = "naive" **then**
2:    Enumerate all distances between each activity
        $a_i \in A$ and each summary path $p_i \in \hat{P}$
3:    $currentGroups \leftarrow$ assign each a_i to the closest p_i
4: **else if** mode = "NOVA" **then**
5:    $V \leftarrow$ virtual node connected to all nodes of all
        $p_i \in \hat{P}$ with zero-weight edges
6:    $Open, T_{nodes} \leftarrow V; Closed, T_{activities} \leftarrow \emptyset$
7:    **repeat**
8:        $n \leftarrow$ closest node in $Open$ to any
        $p_i \in \hat{P}; Closed \leftarrow n$
9:        update T_{nodes} with $x_i.distance$ and $x_i.sp$ for
        each of n's neighbors $x_i \notin Closed$
10:       **if** $x_i \notin Open$ **then** $Open \leftarrow x_i$
11:       **end if**
12:       update $T_{activities}$ with $a_i.distance$ and $a_i.sp$ for
        each activity $a_i \in edge(n, x_i)$
13:       $currentGroups \leftarrow$ assign each a_i to the closest
        p_i based on $T_{activities}$
14:    **until** all active nodes $\in Closed$ or $|Open| = 0$
15:    **if** $\exists$ unassigned activities, a_j **then**
        $currentGroups \leftarrow$ assign each a_j to any p_i
16:    **end if**
17: **end if**
18: **return** $currentGroups$

Fig. 2.6 An example of NOVA. Activity 10 gets assigned to summary path $\langle D, E \rangle$ because the shortest path $\langle V, E, F \rangle$ from V to activity 10 goes through node E of summary path $\langle D, E \rangle$ (Best in color)

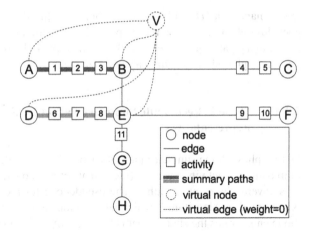

summary path to each activity. It is based on the observation that the shortest path from V to each activity a will go through a node in the summary path that is closest to a.

The NOVA algorithm begins by initializing all the relevant data structures. Line 5 of Algorithm 2 initializes the virtual node V connected to each node of all summary paths by edges of weight 0. Line 6 initializes both the $Open$ list and T_{nodes} to V as well as the $Closed$ and $T_{activities}$ to the empty set. NOVA then expands every node in the $Open$ list based on how close it is to the summary paths; closer nodes get expanded first. Once a node n is expanded, it is moved to the $Closed$ list (line 8). Next, each of $n's$ neighboring nodes $x_i \notin Closed$ is examined, and T_{nodes} is updated with x_i's $distance$ and sp information, where $x_i.distance$ is the network distance of x_i from the nearest summary path, and $x_i.sp$ is x_i's assigned summary path. $x_i.distance$ is calculated by adding $n.distance$ to the distance of edge (n, x_i) (line 9). If x_i is not in $Open$, it is then added to the $Open$ list (line 10).

NOVA records the activity distance to a summary path once it finds activities on an edge connecting node n to the summary path. Every activity a_i that is on edge (n, x_i) is examined, and $T_{activities}$ is updated with a_i's $distance$ and sp information, where $a_i.distance$ is the network distance of a_i from the nearest summary path (based on n), and $a_i.sp$ is the assigned summary path of a_i. Next, an activity is assigned to a summary path (line 12). If the activity was previously assigned to another summary path, it is removed from that path before being assigned to the new summary path. Once all active nodes have been added to the $Closed$ list, or the $Open$ list is empty, NOVA's main loop is stopped (line 13). If unassigned activities remain due to no connectivity to summary paths, these activities are randomly assigned to any summary path $\in \hat{P}$ (line 14). NOVA then terminates and returns the current groups it finds (line 15).

Figure 2.6 shows an example of NOVA activity assignment. Virtual node V is connected by zero weight edges to nodes A and B of summary path $\langle A, B \rangle$ and nodes D and E of summary path $\langle D, E \rangle$ (Algorithm 2, line 5). NOVA assigns activity 10 to summary path $\langle D, E \rangle$ because the shortest path $\langle V, E, F \rangle$ from V to activity 10 goes through node E of summary path $\langle D, E \rangle$. Similarly, NOVA assigns activity 5 to summary path $\langle A, B \rangle$ because the shortest path $\langle V, B, C \rangle$ from V to activity 5 goes through node B of summary path $\langle A, B \rangle$.

2.3.3.2   Phase 2: Recompute Summary Paths (i.e., Selecting Group Representatives)

During phase 2, the summary path of each group is recalculated with the objective of maximizing activity coverage (recall that activity coverage is the number of activities covered by a set of paths). The pseudocode for recomputing summary paths is presented in Algorithm 3, which has two modes naive and D-SPARE. The naive mode enumerates the shortest paths between all active nodes in the spatial network while D-SPARE considers only the set of shortest paths between the active nodes of a group, which gives the correct results.

Performance-Tuning for Phase 2: Performance improvement for phase 2 is achieved via the Divide and Conquer Summary PAth REcomputation (D-SPARE) algorithm, which chooses the summary path for each group with maximum activity coverage ($max Path$) but only considers the set of shortest paths between the nodes of a given group (Algorithm 3, lines 8–9). D-SPARE assumes rich connectivity and otherwise may return a summary path going outside a given fragment. If $max Path$ is null after looking at the shortest paths in a given group, $c_i \in currentGroups$, the shortest path which has the maximum activity coverage based on the activities in c_i is selected as $max Path$ (lines 10–12). $max Path$ is then added to $\hat{P}$ as the new summary path for c_i (line 13). Once all groups $c_i \in currentGroups$ have been considered, $\hat{P}$, which contains the summary paths with maximum activity coverage for each group, is returned.

An example of D-SPARE is shown in Fig. 2.7. The given group consists of activities 1, 2, 3, 4, and 5, and summary path $\langle A, B \rangle$. D-SPARE chooses a new summary path that maximizes activity coverage for every group based on the activities of each group. Summary paths $\langle A, B, C \rangle$ and $\langle C, B, A \rangle$ both maximize activity coverage based on activities 1, 2, 3, 4, and 5, so D-SPARE will choose one of them as the new representative for this group.

KMR with Inactive Node Pruning Performance-Tuning: Inactive node pruning considers only paths between active nodes (vs. paths between all nodes) which reduces the total number of paths considered (Algorithm 1, line 1). An example of inactive node pruning is shown in Fig. 2.1a. The active nodes in this network are $A, B, C, D,$ $E, F,$ and G. Without inactive node pruning, the number of shortest paths considered would be 56 because there are eight nodes, and the shortest path between each node and every other node is considered. With inactive node pruning, the number becomes 42, as only the shortest paths between the seven active nodes are considered.

Algorithm 3 RecomputeSummaryPaths

Input:

    1) a spatial network $G = (N, E)$,

    2) a set of activities A,

    3) a number of routes k,

    4) a set of groups $currentGroups$,

    5) mode $\in \{naive, D\text{-}SPARE\}$

Output:

    k summary paths, $\hat{P}'$, that maximize activity coverage (AC) for each group $\in currentGroups$

Algorithm:

1: **if** mode = "naive" **then**
2:    **for each** $c_i \in currentGroups$ **do**
3:       $P \leftarrow$ shortest paths between active nodes of G
4:       $maxPath \leftarrow$ path in P with Max AC based on
         c_i's activities
5:       $\hat{P}' \leftarrow maxPath$
6:    **end for**
7: **else if** mode = "D-SPARE" **then**
8:    **for each** $c_i \in currentGroups$ **do**
9:       $P' \leftarrow$ the set of shortest paths between the active
         nodes of c_i
10:      $maxPath \leftarrow$ path in P' with Max AC based on
         c_i's activities
11:      **if** $maxPath = \emptyset$ **then**
12:         $P \leftarrow$ shortest paths between active nodes
           of G
13:         $maxPath \leftarrow$ path in P with Max AC based
           on c_i's activities
14:      **end if**
15:      $\hat{P}' \leftarrow maxPath$
16:    **end for**
17: **end if**
18: **return** $\hat{P}'$

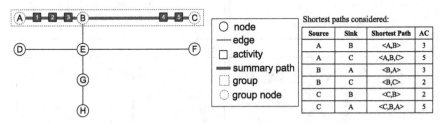

Source	Sink	Shortest Path	AC
A	B	⟨A,B⟩	3
A	C	⟨A,B,C⟩	5
B	A	⟨B,A⟩	3
B	C	⟨B,C⟩	2
C	B	⟨C,B⟩	2
C	A	⟨C,B,A⟩	5

Fig. 2.7 An example of D-SPARE. Only shortest paths between nodes in the group are considered when choosing the new summary path which maximizes activity coverage. In this case, summary paths ⟨A, B, C⟩ and ⟨C, B, A⟩ both maximize activity coverage based on activities 1, 2, 3, 4, and 5 so D-SPARE will choose one of these summary paths as the new representative for this group (Best in color)

2.4 Case Study

A qualitative evaluation comparing KMR with Crimestat K-means [11, 31] on a
real pedestrian fatality data set [32] is shown in Fig. 2.6a. The data consists of 43
pedestrian fatalities (represented as dots) in Orlando, Florida occurring between
2000 and 2009. KMR uses paths and network distance to group activities on a spatial
network whereas in geometry-based techniques such as K-Means, the partitioning
of spatial data is based on grouping similar points distributed in planar space where
the distance is calculated using Euclidean distance. Such techniques focus on the
discovery of the geometry (e.g., circle, ellipse) of high density regions [2] and include
K-means [11, 33–37], K-medoid [12, 13], P-median [14] and Nearest Neighbor
Hierarchical Clustering [15] algorithms.

When evaluating the techniques, it is important to consider both the groups (represented by colors) and the representatives of each group (e.g., paths or ellipses). Pedestrian fatalities usually occur on streets, particularly along arterial roadways [38]. Thus
this activity can be said to have a linear generator. However, the results generated by
Crimestat do not capture this. From Fig. 2.8b, it is clear that the ellipse-based output
is meant for areas, not streets. When Crimestat K-Means is changed to use network
distance, only marginal improvement may be observed. Although the red ellipse in
Fig. 2.8c is aligned with a part of the arterial road, not all the activities on this arterial

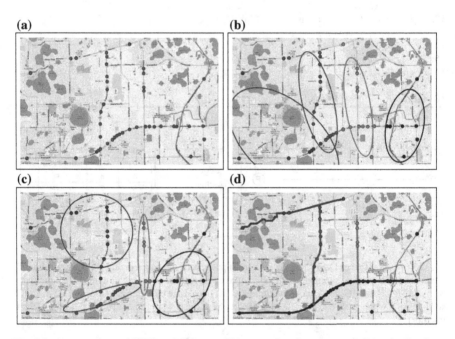

Fig. 2.8 A comparison of KMR and Crimestat K-means when $k = 4$ on pedestrian fatality data
from Orlando, FL [32]. **a** Input. **b** Crimestat K-means with Euclidean Distance. **c** Crimestat K-means
with Network Distance. **d** KMR (Best in color)

are captured. For example, the activities that occur on the road towards the bottom of the figure are split among the red, green, and blue groups. In contrast, the groups of activities in KMR capture the activities on the arterial roads (Fig. 2.8d). For example, the blue group and summary path capture the activities on the arterial road that were split across three groups in network-based K-Means. The group representatives that are paths make sense in this context due to the linear nature of the activities. In another context (in the absence of a linear generator), the geometry-based output of K-Means might make more sense; this is not the case in spatial networks.

2.5 Summary

This chapter explored the challenge of handling a large number of connected components in the spatial network. This challenge was conceptualized as the spatial network activity problem (SNAS), which important application domains such as crime analysis and preventing pedestrian fatalities. The chapter presented the current state-of-the-art techniques, as well as emerging trends such as the KMR algorithm. KMR uses inactive node pruning, Network Voronoi activity Assignment (NOVA) and Divide and conquer Summary PAth REcomputation (D-SPARE) to enhance its performance and scalability. A case study comparing various techniques for addressing SNAS on pedestrian fatality data was presented.

References

1. Oliver, D., Shekhar, S., Kang, J. M., Laubscher, R., Carlan, V., & Bannur, A. (2014). A k-main routes approach to spatial network activity summarization. *IEEE Transactions on Knowledge and Data Engineering, 26*(6), 1464–1478.
2. Eck, J., Chainey, S., Cameron, J., & Wilson, R. (2005). Mapping crime: Understanding hotspots, National Institute of Justice.
3. Matthews, D. A., Effler, S. W., Driscoll, C. T., ODonnell, S. M., & Matthews, C. M. (2008). Electron budgets for the hypolimnion of a recovering urban lake, 1989–2004: Response to changes in organic carbon deposition and availability of electron acceptors". *Limnology and Oceanography, 53*(2), 743–759.
4. Chicago Tribune, Metra argues for delay of 'fail-safe' rail system. https://goo.gl/3bxuw0.
5. Huffington Post, Hungary: Snowstorm strands thousands in their cars. http://www.huffingtonpost.com/huff-wires/20130315/eu-europe-snow.
6. Brantingham, P. J., & Brantingham, P. L. (Eds.) (1981). *Environmental criminology* (pp. 27–54). Beverly Hills: Sage Publications.
7. Levine, N. (2006). Crime mapping and the Crimestat program. *Geographical analysis, 38*(1), 41–56.
8. Scott, M. S., & Dedel, K. (2006). *Assaults in and around bars* (2nd ed.). Washington, DC: Office of Community Oriented Policing Services.
9. Cohen, L. E., & Felson, M. (1979). Social Change and Crime Rate Trends: A Routine. American sociological review (pp. 588–608)
10. Brantingham, P. J., & Brantingham, P. L. (1993). Environment, routine and situation: Toward a pattern theory of crime. *Advances in criminological theory, 5*, 259–294.

11. MacQueen, J. (1967). Some methods for classification and analysis of multivariate observations. *Proceedings of the fifth Berkeley symposium on mathematical statistics and probability*, *1*(14), 281–297.
12. Kaufman, L., & Rousseeuw, P. J. (2009). *Finding groups in data: An introduction to cluster analysis* (Vol. 344). Wiley.
13. Ng, R.T. & Han, J. (1994). Efficient and effective clustering methods for spatial data mining. Proceedings of the International Conference on Very Large Databases.
14. Resende, M. G., & Werneck, R. F. (2004). A hybrid heuristic for the p-median problem. *Journal of Heuristics*, *10*(1), 59–88.
15. D'Andrade, R. G. (1978). U-statistic hierarchical clustering. *Psychometrika*, *43*(1), 59–67.
16. Celik, M., Shekhar, S., George, B., Rogers, J. P., & Shine, J. A. (2007). *Discovering and Quantifying Mean Streets: A Summary of Results, Technical Report 07–025*. Computer Science and Engineering: University of Minnesota.
17. Buchin, K., Cabello, S., Gudmundsson, J., Löffler, M., Luo, J., & Rote, G. et al. (2009). Detecting hotspots in geographic networks, (pp. 217–231). Berlin: Springer.
18. Roach, S. A., & Roach, S. A. (1968). *The theory of random clumping*. London: Methuen.
19. Okabe, A., Okunuki, K. I., & Shiode, S. (2006). SANET: A toolbox for spatial analysis on a network. *Geographical Analysis*, *38*(1), 57–66.
20. Shiode, S., & Okabe, A. (2004). Network variable clumping method for analyzing point patterns on a network. *Unpublished paper presented at the Annual Meeting of the Associations of American Geographers*. Philadelphia, Pennsylvania.
21. Aerts, K., Lathuy, C., Steenberghen, T., & Thomas, I. (2006). Refining spatial clustering of traffic accidents using distances along the network. In *Proceedings of 19th workshop of the international cooperation on theories and concepts in traffic safety*.
22. Spooner, P. G., Lunt, I. D., Okabe, A., & Shiode, S. (2004). Spatial analysis of roadside Acacia populations on a road network using the network K-function. *Landscape Ecology*, *19*(5), 491–499.
23. Steenberghen, T., Dufays, T., Thomas, I., & Flahaut, B. (2004). Intra-urban location and clustering of road accidents using GIS: A Belgian example. *International Journal of Geographical Information Science*, *18*(2), 169–181.
24. Yamada, I., & Thill, J. C. (2007). Local indicators of networkconstrained clusters in spatial point patterns. *Geographical Analysis*, *39*(3), 268–292.
25. Shiode, S., & Shiode, N. (2009). Detection of multiscale clusters in network space. *International Journal of Geographical Information Science*, *23*(1), 75–92.
26. Shekhar, S., & Liu, D. R. (1997). CCAM: A connectivity-clustered access method for networks and network computations. *IEEE Transactions on Knowledge and Data Engineering*, *9*(1), 102–119.
27. Meehan, Bill. (2013). *Modeling electric distribution with GIS*. Redlands: Esri Press.
28. Cormen, T. H. (2001). *Introduction to algorithms*. MIT press.
29. Michael, R. G., & David, S. J. (1979). *Computers and intractability: A guide to the theory of NP-completeness*. San Francisco: W.H. Freeman.
30. Hochbaum, D. S. (1996). Approximating covering and packing problems: set cover, vertex cover, independent set, and related problems. In *Approximation algorithms for NP-hard problems* (pp. 94–143). PWS Publishing Co.
31. Levine, N. (2008). CrimeStat: A spatial statistics program for the analysis of crime incident locations, vol 3.1, Houston, TX: Ned Levine and Associates; and Washington, DC: The National Institute of Justice.
32. Fatality Analysis Reporting System (FARS) Encyclopedia, National Highway Traffic Safety Administration (NHTSA), http://www.nhtsa.gov/FARS.
33. Borah, S., & Ghose, M. K. (2009). Performance analysis of AIM-K-means and K-means in quality cluster generation. ArXiv preprint arXiv:0912.3983.
34. Barakbah, A. R., & Kiyoki, Y. (2009). A pillar algorithm for k-means optimization by distance maximization for initial centroid designation. In *IEEE Symposium on Computational intelligence and data mining, 2009. CIDM'09* (pp. 61–68). IEEE (2009).

35. Khan, S. S., & Ahmad, A. (2004). Cluster center initialization algorithm for K-means clustering. *Pattern recognition letters, 25*(11), 1293–1302.
36. Pelleg, D., & Moore, A. W. (2000). X-means: Extending K-means with efficient estimation of the number of clusters. In *ICML* (Vol. 1).
37. Bradley, P. S., & Fayyad, U. M. (1998). Refining initial points for K-means clustering. *ICML, 98*, 91–99.
38. Ernst, M., Lang, M., & Davis, S. (2011). Dangerous by design: Solving the epidemic of preventable pedestrian deaths, Transportation for America: Surface Transportation Policy Partnership, Washington, DC.

Chapter 3
Many Candidates

Abstract This chapter explores the challenge where the search space itself may have potentially large number of candidates when summarizing spatial network data. For example, in a transportation planning scenario, there may be as many as 10^{16} shortest paths in a given dataset with hundreds of millions of activities or road network nodes. For large roadmaps such as the 100 million road-segments in the US, this results in prohibitive shortest path computation times. This challenge may be formalized as the significant route discovery problem where given a spatial network, a collection of activities (e.g., pedestrian fatality reports, crime reports), and a likelihood threshold θ, the goal is to find all shortest paths in the spatial network where the concentration of activities is unusually high (i.e., statistically significant) and the likelihood exceeds θ. Depending on the domain, an activity may be the location of a pedestrian fatality, a carjacking, a train accident, etc.

3.1 Introduction

The spatial network data summarization challenge of dealing with many candidates may be conceptualized as the problem of Significant Route Discovery (SRD) [1]. SRD identifies routes with significant concentrations of an activity, such as accidents or crimes. Informally, the SRD problem can be defined as follows: given a spatial network, a collection of activities (e.g., pedestrian fatality reports, crime reports), and a likelihood threshold θ, find all shortest paths in the spatial network where the concentration of activities is unusually high (i.e., statistically significant) and the likelihood exceeds θ. Depending on the domain, an activity may be the location of a pedestrian fatality, a carjacking, a train accident, etc. Figure 3.1a, b illustrate an input and output example of SRD, respectively. The input consists of seven nodes, seven edges (with edge weights set to 1 for illustration purposes, shown as the second number on each edge), twenty activities (shown as the first number in red on each edge), and $\theta = 2$, indicating that we are interested in shortest paths whose likelihood exceeds $\theta = 2$. The output contains two shortest paths, $\langle N_1, N_2, N_3 \rangle$ and $\langle N_6, N_5, N_7 \rangle$ that are at least twice as likely to have instances of the activity (e.g., accidents, crime).

© The Author(s) 2016
D. Oliver, *Spatial Network Data*, SpringerBriefs in Computer Science,
DOI 10.1007/978-3-319-39621-7_3

(a)

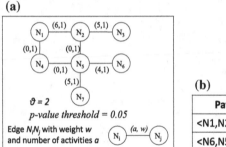

(b)

Path	λ	P-value
<N1,N2,N3>	3.05	0.007
<N6,N5,N7>	2.04	0.026

Fig. 3.1 Example of significant route discovery. Activity information is aggregated by edge (Best in color). **a** Input. **b** Output

3.1.1 Challenges

SRD is challenging due to the potentially large number of candidate routes ($\sim 10^{16}$) in a given dataset with millions of activities or road network nodes. For large roadmaps such as the 100 million road-segments in the US, this results in prohibitive short-est path computation times. Additionally, significance testing does not obey the monotonicity property, meaning that there is no ordering between the likelihood of a path and its super-paths, or vice-versa. In other metrics such as activity count, for example, a path will always have less than or equal to the number of activities of its super-paths, a property which may be exploited for computational speedup. How-ever, this property does not hold for significance testing. Furthermore, depending on the method used to determine statistical significance, computation times may also be impacted (e.g., $m = 1000$ Monte Carlo simulations may be required to calculate statistical significance).

3.1.2 Current State-of-the-Art

Dividing spatial data into statistically significant groups is an important task in many domains (e.g., transportation planning, public health, epidemiology, climate science, etc.) [2]. Methods for this type of partitioning may generally be considered to be geometry-based or network-based.

Geometry-based techniques [3–5] partition spatial data using two-dimensional shapes (e.g., circles, rectangles). This is useful in domains such as public health, where finding spatial clusters with a higher density of disease is of interest for under-standing the distribution and spread of diseases, outbreak detection, etc. Kulldorff et al. proposed a spatial scan statistics framework (and the SaTScan software) for dis-ease outbreak detection [3]. The spatial scan statistic employs a likelihood ratio test where the null hypothesis is the probability that disease inside a region is the same as outside, and the alternate hypothesis is that there is a higher probability of disease inside than outside. All the spatial regions, represented by a circle or rectangle in the spatial framework, are enumerated and the one that maximizes the likelihood ratio is identified as a candidate. However, if we apply SaTScan to a road network, many significant routes may be missed since a large fraction of the area bounded by circles for activities on a path will be empty, as shown in Fig. 3.6b. Furthermore, geometry-based techniques may not be appropriate for modeling linear clusters, which are formed when the underlying generator of the phenomena is inherently linear (e.g., pedestrian fatalities, railroad accidents, etc.).

Network-based techniques [6–9], on the other hand, leverage the underlying spatial network when partitioning spatial data. For example, Linear Intersecting Paths (LIP) [8] and Constrained Minimum Spanning Trees (CMST) [6] utilize a subgraph (e.g., a path or tree) to discover statistically significant groups.

In LIP [8], one anomalous sub-component out of a set of connected paths that intersect each other is discovered. The connected paths are based on locations in the spatial network with the highest percentage of activities, specified by the user. Hence the likelihood ratio is only evaluated on a portion of the graph specified by this percentage, not the entire spatial network. Figure 3.2 shows an example input and output of LIP. The user-specified percentage is 30%, which means all the candidates will have paths containing edge $\langle N_1, N_2 \rangle$ since this edge has six activities (out of a possible 20 activities). Examples of possible candidates are $\langle N_1, N_2, N_3 \rangle$, $\langle N_1, N_2, N_5 \rangle$, $\langle N_2, N_1, N_4 \rangle$, $\langle N_1, N_2, N_5, N_7 \rangle$, etc. The output is $\langle N_1, N_2, N_3 \rangle$, since it has the highest likelihood (Sect. 3.2 details how the likelihood ratio is calculated). However, in addition to returning only one statistically significant component, the results of this approach are sensitive to the percentage of the network selected. If the percentage is too high, the number of candidates may be highly restricted, which could result in not identifying statistically significant regions of interest. If the percentage is too low, LIP may be computationally prohibitive due to the large number of candidates.

Another network-based technique, CMST [6], finds one statistically significant tree in the spatial network. Figure 3.2c shows an example of this approach. Here the

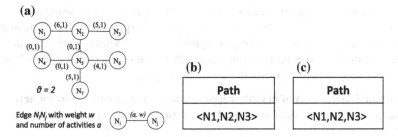

Fig. 3.2 Simplified example, **a** input, where activities are summarized by total count per edge, **b** output of Linear Intersecting Paths (LIP) [8], and **c** output of Constrained Minimum Spanning Trees (CMST) [6] (Best in color)

output is $\langle N_1, N_2, N_3 \rangle$, since this tree has the highest likelihood. However, in addition to returning only one statistically significant tree, the size of the tree is restricted, which could result in not identifying statistically significant regions of interest.

3.1.3   Outline of the Chapter

The chapter is organized as follows: Sect. 3.2 presents the basic concepts and problem statement of SRD. Section 3.3 presents recent trends towards addressing SRD. Section 3.4 presents a case study comparing a significant network-based output (i.e., shortest paths) to a significant geometry-based output (e.g., circles) on pedestrian fatality data. Section 3.5 presents a discussion outlining potential future directions and Sect. 3.6 concludes the chapter.

3.2   Basic Concepts and Problem Statement

This section reviews several key concepts in SRD and presents a formal problem statement.

3.2.1   Basic Concepts

The basic concepts are defined as follows:

Definition 7 A **spatial network** $G = (N, E)$ consists of a node set N and an edge set E, where each element u in N is associated with a pair of real numbers (x, y) representing the spatial location of the node in a Euclidean plane [10]. Edge set E is a subset of the cross product $N \times N$. Each element $e = (u, v)$ in E is an edge that joins node u to node v.

Figure 3.1a shows an example of a spatial network where circles represent nodes and lines represent edges. A road network is an example of a spatial network where nodes represent street intersections and edges represent streets. In utility networks such as electric networks, nodes represent devices such as transformers and fuses whereas edges represent lines (e.g., medium voltage lines) [11].

Definition 8 AR is a **set of activity reports**, where each activity report (or **activity**) ar_i has a location on an edge $e = (u, v)$. $a(e = (u, v))$ is the number of activities on an edge $e = (u, v)$.

In transportation planning, an activity may be the location of a pedestrian fatality; in crime analysis, an activity may be the location of a theft. Each edge in Fig. 3.1a is associated with a number of activities (e.g., edge $\langle N_1, N_2 \rangle$ has 6 activities).

Definition 9 The **activity coverage** *inside* **a path**, a_p, is the number of activities on p. The **activity coverage** *outside* p is $|AR| - a_p$, where $|AR|$ is the total number of activities in the spatial network, G.

For example, in Fig. 3.1a, the activity coverage *inside* path $\langle N_1, N_2, N_3 \rangle$ is 11 whereas the activity coverage *outside* $\langle N_1, N_2, N_3 \rangle$ is $20 - 11 = 9$.

Definition 10 The **weight** *inside* **a path**, w_p, is the sum of weights of all edges in p. The **weight** *outside* p is $|W| - w_p$, where $|W|$ is sum of weights of all edges in G.

In Fig. 3.1a, the weight *inside* $\langle N_1, N_2, N_3 \rangle$ is 2 whereas the weight *outside* $\langle N_1, N_2, N_3 \rangle$ is $7 - 2 = 5$.

Definition 11 The **likelihood ratio of path** p, $\lambda_p = \frac{a_p \div w_p}{(|AR| - a_p) \div (|W| - w_p)}$ [3, 9].

The likelihood ratio of path p, λ_p is the ratio of the activity density *inside* path p to the activity density *outside* p. Activity density may be estimated in different ways across different domains. In transportation planning, activity density inside p may be estimated using $\frac{a_p}{VMT}$, where VMT is vehicle miles traveled (i.e., the total number of miles driven by all vehicles within a given time period and geographic area). Path weight may also be used to estimate activity density [9]. In Fig. 3.1a, $\lambda_{\langle N_1, N_2, N_3 \rangle} = \frac{11 \div 2}{9 \div 5} = 3.05$.

Definition 12 A **super-path** of path p is any path sp that contains p, where sp is a subset of G. A **sub-path** is a path making up part of the super-path.

For example, in Fig. 3.1a, $\langle N_1, N_2, N_5, N_6 \rangle$ and $\langle N_1, N_2, N_5, N_7 \rangle$ are super-paths of $\langle N_1, N_2, N_5 \rangle$. Conversely, $\langle N_1, N_2, N_5 \rangle$ is a sub-path of $\langle N_1, N_2, N_5, N_6 \rangle$.

3.2.2 Problem Statement

The problem of Significant Routes Discovery (SRD) can be expressed as follows:
Given:

1. A spatial network $G = (N, E)$ with a set of activities with point locations on network nodes or edges and weight function $w(u, v) > 0$ for each edge $e = (u, v) \in E$ (e.g., network distance),
2. A likelihood ratio (λ) threshold, θ,
3. A p-value,
4. m, indicating the number of Monte Carlo simulations

Find: All routes $r \in R$ with likelihood ratio $\lambda_r \geq \theta$ and a p-value significance level
Objective: Computational efficiency

Constraints:

1. Each route $r \in R$ is a shortest path between its end-nodes,
2. $r_i \in R$ is not a subset of any $r_j \in R \; \forall r_i, r_j \in R$ where $r_i \neq r_j$,
3. Correctness and completeness

The spatial network input for SRD is defined in Definition 7. The θ input is a threshold indicating the minimum desired likelihood ratio. The p-value input is the desired level of statistical significance and m indicates the number of Monte Carlo simulations for determining statistical significance. The output for SRD is a set of shortest paths meeting the desired likelihood ratio and level of statistical significance. The shortest paths returned are constrained so that they are not sub-paths of any other path in the output. This constraint aims to improve solution quality by reducing redundancy in the paths returned.

Example. The network in Fig. 3.1a can be viewed as a road network, composed of streets (edges) and intersections (nodes). The aim is to find significant shortest paths that meet the given likelihood threshold of 2. In other words, find shortest paths that are twice as likely to have pedestrian fatalities. In a transportation planning scenario, identifying such routes would guide street redesign efforts to reduce the risk of pedestrian fatalities (e.g., adding sidewalks, crosswalks, pedestrian refuges, street lighting, etc.). In Fig. 3.1b, routes $\langle N_1, N_2, N_3 \rangle$ and $\langle N_6, N_5, N_7 \rangle$ are returned since they are shortest paths whose likelihood exceeds $\theta = 2$ and they are not sub-paths of any other path in the output.

3.2.2.1   Finding Significant Paths

Each shortest path in the spatial network is evaluated for statistical significance using Monte Carlo simulations to determine whether or not it is truly anomalous. Here the null hypothesis states that the paths identified by the path likelihood ratio are random or by chance alone. The likelihood ratio is supplemented with a p-value to decide whether the null hypothesis should be rejected in the hypothesis test. The p-value is the probability of obtaining a value of a given likelihood ratio as equally or more extreme than that observed by chance alone.

In the Monte Carlo simulations, each activity in the original graph G is randomly associated with an edge so that the number of activities on each edge is shuffled, forming a new graph G_s. Note that all the activities in G are present in G_s, with no activities added or removed; the original activities in G are now shuffled so they may be on different edges in G_s. We then compare the highest likelihood threshold λ_{maxG_s} of randomized G_s with the highest λ_{maxG} of original G. If the original one is smaller (i.e., $\lambda_{maxG} < \lambda_{maxG_s}$), then $p = p + 1$. The above process repeats m times and after it terminates, the p-value is subsequently p/m. Paths whose p-values are less than or equal to the given p-value threshold are deemed statistically significant.

3.3 Trends

An initial (albeit naïve) solution to the SRD problem as well as an enhanced version is described in this section.

3.3.1 Naïve Significant Route Miner (NaïveSRM)

Algorithm 4 presents the pseudocode for the NaïveSRM approach. The basic idea behind the algorithm is to find all statistically significant shortest paths in the spatial network whose likelihood exceeds θ, under the constraint that the shortest paths returned are not sub-paths of any other path in the output. Algorithm 4 proceeds by calculating all shortest paths, P, in the spatial network (Line 1). Line 2 evaluates each shortest path in P to determine if it meets the given likelihood threshold θ to form a *Candidates* set. In line 3, the statistical significance of each shortest path in *Candidates* is evaluated and the significant routes are stored in *SigRoutes*. In order to assess statistical significance, all shortest paths in each of the m simulated graphs are used to calculate the p-value. In line 4, all paths in *SigRoutes* that are not sub-paths of any other path in *SigRoutes* are returned, and the algorithm terminates. The purpose of returning significant routes that are not sub-paths of any other path is to improve solution quality. For example, if $\langle N_1, N_2 \rangle$ and $\langle N_1, N_2, N_3 \rangle$ are both found to be significant, only $\langle N_1, N_2, N_3 \rangle$ is returned.

Algorithm 4 Naïve Significant Route Miner (NaïveSRM) Algorithm

---

```
Input:
    1) A spatial network G = (N, E) with a set of activities with point locations
    on network nodes or edges and weight function w(u, v) > 0 for each edge
    e = (u, v) ∈ E (e.g., network distance),
    2) A likelihood ratio (λ) threshold, θ,
    3) A p-value threshold,
    4) m, indicating the number of Monte Carlo simulations
Output:
    All routes r ∈ R with λ_r ≥ θ and p-value significance level
Algorithm:
1: {Step 1:} P ← calculate all-pairs shortest path in G
2: {Step 2:} Candidates ← paths in P having λ ≥ θ
3: {Step 3:} SigRoutes ← significant paths in Candidates using m Monte Carlo
    simulations
4: {Step 4:} return paths that are not sub-paths of any other path in SigRoutes
```

NaïveSRM Example: Figure 3.3 shows an example execution trace of NaïveSRM. The spatial network has 7 nodes, 7 edges, and 20 activities, represented by the first number in red on each edge (e.g., edge $\langle N_1, N_2 \rangle$ has six activities). The given likelihood ratio threshold θ is set to 2 and the p-value is set to 0.05.

In step 1 of Fig. 3.3, all shortest paths in the given spatial network are calculated. For example, the shortest path between nodes N_1 and N_3 is $\langle N_1, N_2, N_3 \rangle$. Next, in step 2, the likelihood ratio, λ, for each shortest path is determined (see Definition 11)

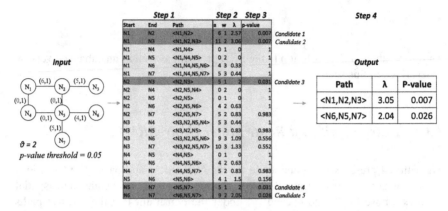

Fig. 3.3 Execution trace of Naïve Significant Route Miner (NaïveSRM). *Circles* represent nodes and *lines* represent edges (Best in color)

and those whose $\lambda \geq \theta$ are stored as candidate solutions. In the figure, the five highlighted paths $\langle N_1, N_2 \rangle$, $\langle N_1, N_2, N_3 \rangle$, $\langle N_2, N_3 \rangle$, $\langle N_5, N_7 \rangle$, and $\langle N_6, N_5, N_7 \rangle$ are all candidates since their likelihood ratios meet or exceed the threshold of 2. In step 3, the statistical significance of each candidate is calculated using Monte Carlo simulations (discussed next). All five candidates meet the p-value threshold of 0.05. In step 4, the paths among significant paths that are not sub-paths of any other path are returned. In this example, paths $\langle N_1, N_2, N_3 \rangle$ and $\langle N_6, N_5, N_7 \rangle$ are returned. Paths $\langle N_1, N_2 \rangle$, $\langle N_2, N_3 \rangle$, and $\langle N_5, N_7 \rangle$ were not returned (even though they met and exceeded the likelihood and p-value thresholds) because they are each sub-paths of the two paths that were returned.

3.3.2 Significant Route Miner with Likelihood Pruning and Monte Carlo Speedup (SRM)

Algorithm 5 presents the pseudocode of the SRM [1] approach. SRM uses filter and refine techniques (e.g., Likelihood Ratio pruning and Monte Carlo speedup) to achieve computational savings. Filter and refine techniques may not change worst case complexity but they can reduce runtime in many cases. Likelihood Ratio pruning creates a boundary via the upperbound likelihood ratio such that not all destinations are visited from each source node. Some of the destinations are pruned because the shortest paths to them will never meet the likelihood ratio threshold. Monte Carlo speedup avoids generating all shortest paths in cases where a shortest path in the simulated dataset has a higher likelihood ratio than the shortest paths in the original dataset. Monte Carlo speedup also terminates early if the p-value threshold will not be met based on the number of times the maximum likelihood ratio in the simulated dataset beats the maximum likelihood ratio in the original dataset.

3.3.2.1 Likelihood Pruning

Likelihood pruning aims to avoid calculating all shortest paths in G based on the given threshold θ. It is based on the idea that for each shortest path p, it is possible to determine an upper bound likelihood ratio for the super-paths rooted at p's start node, without calculating those super-paths.

Definition 13 The **upperbound likelihood ratio for path** p,
$\hat{\lambda}_p = \frac{\hat{a}_p \div \hat{w}_p}{(|AR| - \hat{a}_p) \div (|W| - \hat{w}_p)}$, where $\hat{a}_p = a_p + (|AR| - a_t)$ (where a_t is the number of activities in the shortest path tree rooted at p's source node) and $\hat{w}_p$ is the weight of the shortest super-path of p, rooted at p's start node.

The intuition behind the upper bound likelihood ratio for path p is that (1) the number of activities on all of p's super-paths rooted at p's start node are bounded by the number of activities in the spatial network minus the number of activities in the current shortest path tree rooted at the source node in p and (2) the weight of any super-path of p is at least the weight of the closest edge to p plus p's weight.

Lines 1–12 of Algorithm 5 shows the pseudocode for likelihood pruning, which is similar to Dijkstra's algorithm [12] with a few exceptions: (1) the shortest paths from a single active node to all destinations are calculated for all active nodes in the spatial network, (2) if the upper bound likelihood ratio for path $\langle s...u \rangle$ is below the given likelihood threshold θ, u's neighbors are not visited (line 6), and (3) upperbound statistics are calculated and updated each time the weight from source s to a node v is updated (lines 9–12).

Likelihood Pruning Example: Figure 3.4a illustrates the basic idea behind likelihood pruning. In this example, we have set the likelihood threshold to $\theta = 5$, indicating that we are interested in paths that are five times as likely to have pedestrian fatalities. During the algorithm's execution, at some point the source node becomes N_1, and the shortest path between N_1 and every other active node in the spatial network is calculated. When the shortest path between N_1 and N_5 is calculated, the upper bound likelihood ratio for path $\langle N_1, N_2, N_5 \rangle$ is determined to be 4, since based on Definition 13, the calculation would be $\frac{(6+(20-11)) \div 3}{(20-((6+(20-11)) \div (7-3))}$, where

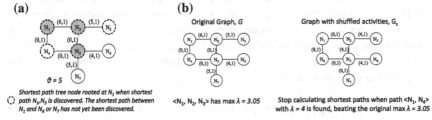

(a)

$\vartheta = 5$

Shortest path tree node rooted at N_1 when shortest
path N_3,N_5 is discovered. The shortest path between
N_1 and N_6 or N_7 has not yet been discovered.

(b)

Original Graph, G

$\langle N_1, N_2, N_3 \rangle$ has max $\lambda = 3.05$

Graph with shuffled activities, G_s

Stop calculating shortest paths when path $\langle N_1, N_4 \rangle$
with $\lambda = 4$ is found, beating the original max $\lambda = 3.05$

Fig. 3.4 **a** Example of Likelihood Pruning. Since we know the upper-bound likelihood for $\langle N_1, N_2, N_5 \rangle$ is 4, we can avoid calculating the shortest paths $\langle N_1, N_2, N_5, N_6 \rangle$ and $\langle N_1, N_2, N_5, N_7 \rangle$ for $\theta = 5$. **b** Example of Monte Carlo Speedup (Best in color)

Algorithm 5 Significant Route Miner with Likelihood Pruning and Monte Carlo Speedup (SRM) Algorithm

---

```
Inputs and Outputs for SRM are same as NaïveSRM
Algorithm:
    {Step 1: Likelihood Pruning}
 1: for each s ∈ active nodes in G do
 2:     Initialize D[v] ← inf; Pred[v] ← ∅; Λ̂[v] ← θ; a[v] ← 0; aₜ ← 0; D[s] ← 0; PQ ← N
 3:     while PQ ≠ ∅ do
 4:         u ← node in PQ with smallest distance in D[]; P ← shortest path (s, u)
    in Pred[]
 5:         aₜ ← aₜ+ number of activities on edge Pred[u]
 6:         if Λ̂[v] ≥ θ then
 7:             for each v adjacent to u do
 8:                 sum ← D[u] + w(u, v)
 9:                 if sum < D[v] then
10:                     D[v] ← sum; update v's position in PQ based on sum; Pred[v] ← u
11:                     a[v] ← a[u] + a(u, v); ŵ ← sum+ weight of closest neighbor w(u, v)
12:                     Λ̂[v] ← calculate λ̂ₛᵥ based on a[v], aₜ and ŵ
13: {Step 2:} Candidates ← paths in P having λ ≥ θ
    {Step 3: Monte Carlo Speedup}
14: λₘₐₓ𝒢 ← highest likelihood ratio in G
15: for each simulation₁....simulationₘ do
16:     Gₛ ← assign activities in G to random edges
17:     λₘₐₓ𝒢ₛᵢ ← 0
18:     for each shortest path p ∈ Gₛ do
19:         if λₚ > λₘₐₓ𝒢ₛᵢ then
20:             λₘₐₓ𝒢ₛᵢ ← λₚ; pₘₐₓᵣ ← pₘₐₓᵣ + 1
21:             if pₘₐₓᵣ/N ≤ p-value threshold then return ∅
22:             if λₚ > λₘₐₓ𝒢 then break
23:     for each route r ∈ Candidates do
24:         if λₘₐₓ𝒢ₛᵢ > λₘₐₓ𝒢 then pᵣ ← pᵣ + 1
25: for each route r ∈ Candidates do
26:     if pᵣ/N ≤ p-value threshold then SigRoutes ← r
27: {Step 4:} return paths that are not sub-paths of any other path in SigRoutes
```

$\hat{a}_p = 6 + (20 - 11) = 15$ and $\hat{w}_p = 2 + 1 = 3$. We can, therefore, avoid calculating the shortest paths $\langle N_1, N_2, N_5, N_6 \rangle$ and $\langle N_1, N_2, N_5, N_7 \rangle$ for $\theta = 5$.

Lemma 3.1 $\hat{a}_p = a_p + (|AR| - a_t)$ *is a correct upper bound for the number of activities of all super-paths of p, rooted at p's source node.*

Proof Consider the most current shortest path tree, t, generated when the shortest path p from node u to node v is found in the spatial network. By definition, p is unique in t. Thus, all super-paths of p, rooted at u will not contain any nodes or edges in t since t was generated when v was discovered (closed). Therefore, super-paths of p rooted at u cannot contain any activities already in t, except those activities already in p. Hence, $\hat{a}_p = a_p + (|AR| - a_t)$ is a correct upper bound for the number of activities of p, rooted at p's source node.

Lemma 3.2 $\hat{\lambda}_p = \frac{\hat{a}_p \div \hat{w}_p}{(|AR| - \hat{a}_p) \div (|W| - \hat{w}_p)}$ *is a correct upper bound on the likelihood ratio for any super-path of p, rooted at p's source node, where a_t is the number of activities in the shortest path tree rooted at p's source node and $\hat{w}_p$ is the weight of the shortest super-path of p, rooted at p's start node.*

Proof $\hat{a}_p = a_p + (|AR| - a_t)$ is a correct upper bound for the number of activities of all super-paths of p, rooted at p's source node (Lemma 3.1). Since $\hat{w}_p$ is the weight of the shortest super-path of p rooted at p's source node, $\hat{\lambda}_p$'s value is maximized (based on its location in the denominators in Definition 13).

Theorem 3.1 *Likelihood pruning is a correct pruning method*

Proof A pruning method is considered correct if it does not exclude any optimal solutions (i.e., shortest paths whose likelihood ratio exceeds θ). By Lemma 3.2, $\hat{\lambda}_p$ is a correct upper bound on the likelihood ratio for any super-path of p rooted at p's source node. Since likelihood pruning does not prune out any path that meets or exceeds the upperbound likelihood ratio, and hence the likelihood ratio, no such optimal solutions will be pruned. Thus, likelihood pruning is correct.

3.3.2.2 Monte Carlo Speedup

Monte Carlo speedup aims to calculate the p-value without considering all shortest paths in each simulated graph. The basic idea is that once a shortest path in the simulated graph is found to have a higher likelihood ratio than the maximum likelihood ratio in the original graph, the simulation immediately ends with the p-value being incremented. In other words, there is no reason to keep looking at all shortest paths in the simulated graph if we find one that already beats the maximum likelihood ratio in the original graph. Additionally, Monte Carlo speedup stops all simulations the moment p out of m simulations are found where the simulated likelihood ratio beats the original maximum likelihood ratio. In other words, there is no reason to execute all m simulations if we find that the p-value threshold will not be met. The pseudocode for Monte Carlo speedup is presented in Lines 14–26 of Algorithm 5.

Monte Carlo Speedup Example: Figure 3.4b illustrates one of the basic ideas behind Monte Carlo speedup. In this example, the graph on the left is the original graph G whereas the graph on the right, G_s, represents one simulation with the activities shuffled. In G_s, instead of looking at all 42 shortest paths, we can stop and increment p the moment a path that has a likelihood higher than the maximum likelihood in G is found. In this case, that path would be $\langle N_1, N_4 \rangle$ (on the right of the figure), with a likelihood ratio of 4.

Theorem 3.2 *Monte Carlo Speedup is a correct method for calculating p-value.*

Proof A method for calculating p-value is correct if it accurately determines the number of times, p, out of m simulations that the simulated likelihood ratio beats the original maximum likelihood ratio. Monte Carlo speedup only increments p when a likelihood ratio is found in the simulated graph that beats the original likelihood ratio. Thus, Monte Carlo Speedup is correct.

3.3.3   *Dynamic Segmentation*

A fundamental challenge that many network-based techniques face is dealing with very long edges with a dense cluster of activities on one end (e.g., edge $\langle N_1, N_2 \rangle$ in Fig. 3.1a). Long empty portions of these edges reduce the likelihood ratio of the entire edge and such edges are often excluded by previous approaches even though many users may prefer to include the portions with dense clusters of activities. In other words, many approaches lack the capability to consider paths between activity locations and instead only focus on paths between network nodes. Consequently, they may miss paths containing activities that are close to each other but are on a long edge.

Figure 3.5 shows an example of a spatial network that has been dynamically segmented. Dynamic segmentation facilitates the inclusion of paths with dense portions of activities such as $\langle A1, A2, A3, A4, A5, A6, A7 \rangle$, which has a higher likelihood ratio (14) than that of the enclosing path $\langle N_1, N_2, N_3 \rangle$ (5.83) in the traditional network.

Resolving statistically significant routes to the sub-edge level requires a dynamic segmentation data model. In SRM's dynamic segmentation model, the network structure is altered such that new nodes are formed at the locations of activities and new edges are added to connect these nodes. Dynamic segmentation enables us to evaluate paths that start and end with activities, which may be in the middle of an edge in the original spatial network. As such, segments which were previously not tested for statistical significance or which may have been previously deemed "not significant" because they were on a long edge, may end up as part of the result.

Fig. 3.5 Example of dynamic segmentation where paths between activities at the sub-edge level are considered

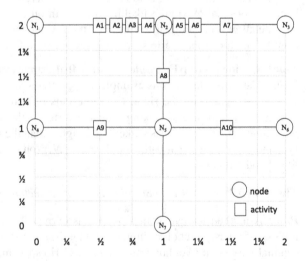

3.4 Case Study

This section presents a qualitative evaluation of the significant route miner (SRM) with SaTScan [13] (continuous Poisson process) on a real pedestrian fatality data set [14], shown in Fig. 3.6a. As noted earlier, SaTScan discovers areas of significant activity that are represented as circles on the spatial network while SRM discovers significant shortest paths. The input consisted of 43 pedestrian fatalities (represented as dots) in Orlando, Florida occurring between 2000 and 2009. For each edge (portion of road) in the network, fatality count was aggregated, yielding overall activity, and weight was the actual road network distance. The maps were prepared using QGIS' Open Layers plugin [15], and the road network was from the US Census Bureaus TIGER/Line Shapefiles [16].

When evaluating the techniques, the outputs of circles versus shortest paths are considered. While p-value thresholds of 0.05 or lower are often desired, we used a p-value threshold of 0.15 because the circles chosen by SaTScan had high p-values for this dataset. As noted earlier, pedestrian fatalities usually occur on streets, particularly along arterial roadways [17]. Thus this activity can be said to have a linear generator. However, the results generated by SaTScan do not capture this. From Fig. 3.6b, it

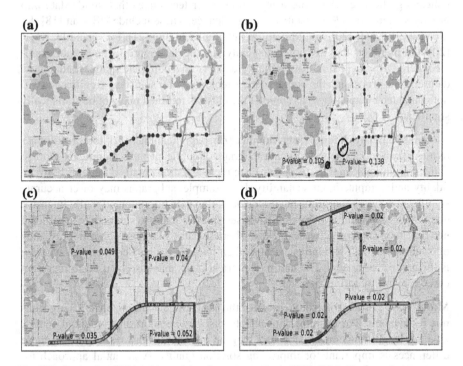

Fig. 3.6 Comparing SRM (without dynamic segmentation), SRM_DS (with dynamic segmentation), and SaTScan's output for a p-value threshold of 0.15 and $\theta = 1.75$ on pedestrian fatality data from Orlando, FL [14] (Best in color). **a** Input. **b** SaTScan. **c** SRM. **d** SRM_DS

is clear that the circle-based output is meant for areas, not streets. In contrast, the shortest paths detected by SRM fully capture the significant activities on the arterial roads (some of the paths in Fig. 3.6c are overlapping). Furthermore, the paths in the figure make sense in this context due to the inherently linear nature of the activities.

The output of SRM with dynamic segmentation (SRM_DS) was also compared to the outputs of SaTScan and SRM (miner without dynamic segmentation). As we can observe from Fig. 3.6d, relative to SaTScan, SRM_DS is able to capture significant activities on the arterial roads (just as SRM). In contrast to SRM, SRM_DS is also able to detect paths occurring on the sub-edge level such as the blue path (top-center of the figure). Therefore, even if there were paths that were not significant because the activities on them were on long edges, if the activities were close to each other on the network, they may show up in SRM_DS's result.

3.5   Discussion

Techniques without significance testing: In addition to the challenge of many candidates, another focus of this chapter was partitioning techniques that consider statistical significance. There are a myriad of other techniques that divide data into groups without considering statistical significance. These include DBScan [18], K-Means [19], KMR [20], and Maximum Subgraph Finding [21]. For example, KMR [20] returns routes that are not statistically significant. Figure 3.7 shows an example where DBScan [18] returns 7 chance clusters on a complete spatial randomness dataset. Post-processing the output of these techniques for statistical significance will not guarantee completeness as some of the clusters returned may not be statistically significant.

Alternative network footprints: Summarizing significant network footprints of activities may be done using significant subgraphs, significant paths, significant shortest paths, minimum spanning trees, etc. Each representation entails a tradeoff between fidelity and computational scalability. For example, subgraphs may offer accurate significant network footprints but their calculation may be computationally intensive due to their exponential number. SRM uses shortest paths to summarize significant network footprints of activities. While shortest paths may lose some fidelity, they offer computational scalability because their number is bounded (i.e., $\binom{n}{2}$, where n is the number of nodes). The union of shortest paths may also be used to represent other network footprints.

Multi-Scale Model: Different regions of the network may require different algorithmic refinements. For example, linear hotspots may be short on residential streets, medium length on county roads, and long on interstate highways. Modeling these differences is important for improving solution quality. A potential approach towards this end is a multi-scale model, since roads in transportation networks are often categorized into highways, county roads, residential streets, etc.

Fig. 3.7 *Colored dots* are part of chance clusters identified by DBScan [18] on a complete spatial randomness dataset (Best in color)

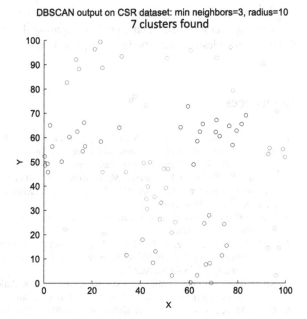

DBSCAN output on CSR dataset: min neighbors=3, radius=10
7 clusters found

Alternate Problem Formulations: In an alternate formulation of the problem, the spatial network may be modeled with an activity count function $a(u) \geq 0$ for each node. The idea is that activities may also occur at nodes in addition to being distributed within network edges. In this way, the current approach may be extended to capture activities at nodes (e.g., vehicle accidents). If activities are modeled as counts at each node, this may alter the computational structure.

In another formulation of the problem, one may require every edge with significant linear hotspots to meet certain criteria (e.g., minimum likelihood ratios and p-values). This alternate problem formulation may exhibit a different computational structure (e.g., monotonicity) and allow more aggressive bottom-up pruning that is popular in the graph mining literature. Here the results may be significantly different from those in the current problem formulation.

3.6 Summary

This chapter explored the challenge of a large search space with many candidates. This challenge was formalized as the problem of significant route discovery in relation to important application domains such as preventing pedestrian fatalities and crime analysis. The chapter presented a significant route miner that discovers multiple statistically significant shortest paths in a spatial network. This approach uses Likelihood Pruning, Monte Carlo Speedup, and Dynamic Segmentation to enhance its performance and scalability, which is important when dealing with many candidates in a spatial network. A case study comparing the significant route miner with SaTScan on pedestrian fatality data was presented.

References

1. Oliver, D., Shekhar, S., Zhou, X., Eftelioglu, E., Evans, M. R., & Zhuang, Q., et al. (2001). Significant route discovery: A summary of results. In *Geographic information science* (pp. 284–300). Berlin: Springer International Publishing.
2. Shekhar, S., Evans, M. R., Kang, J. M., & Mohan, P. (2011). Identifying patterns in spatial information: A survey of methods. *Wiley Interdisciplinary Reviews: Data Mining and Knowledge Discovery, 1*(3), 193–214.
3. Kulldorff, M. (1997). A spatial scan statistic. *Communications in Statistics-Theory and methods, 26*(6), 1481–1496.
4. Neill, D. B., & Moore, A. W. (2004). Rapid detection of significant spatial clusters. In *Proceedings of the tenth ACM SIGKDD international conference on Knowledge discovery and data mining* (pp. 256–265). ACM.
5. Kulldorff, M. (1999). Spatial scan statistics: models, calculations, and applications. In *Scan statistics and applications* (pp. 303–322). Birkhuser Boston.
6. Costa, M. A., Assuno, R. M., & Kulldorff, M. (2012). Constrained spanning tree algorithms for irregularly-shaped spatial clustering. *Computational Statistics and Data Analysis, 56*(6), 1771–1783.
7. Duczmal, L., & Assuncao, R. (2004). A simulated annealing strategy for the detection of arbitrarily shaped spatial clusters. *Computational Statistics and Data Analysis, 45*(2), 269–286.
8. Shi, L., & Janeja, V. P. (2011). Anomalous window discovery for linear intersecting paths. *IEEE Transactions on Knowledge and Data Engineering, 23*(12), 1857–1871.
9. Janeja, V. P., & Atluri, V. (2005). LS 3: A linear semantic scan statistic technique for detecting anomalous windows. In *Proceedings of the 2005 ACM symposium on applied computing* (pp. 493-497). ACM.
10. Shekhar, S., & Liu, D. R. (1997). CCAM: A connectivity-clustered access method for networks and network computations. *IEEE Transactions on Knowledge and Data Engineering, 9*(1), 102–119.
11. Meehan, Bill. (2013). *Modeling electric distribution with GIS.* Redlands: Esri Press.
12. Cormen, T. H. ((2009)). *Introduction to algorithms.* MIT press.
13. Kulldorff, M., Rand, K., Gherman, G., Williams, G., & DeFrancesco, D. (1998). SaTScan v 2.1: Software for the spatial and space-time scan statistics. Bethesda, MD: National Cancer Institute.
14. Fatality Analysis Reporting System (FARS) Encyclopedia, National Highway Traffic Safety Administration (NHTSA), http://www.nhtsa.gov/FARS
15. Quantum GIS OpenLayers Plugin. Retrieved January 23, 2014. http://plugins.qgis.org/plugins/openlayers_plugin
16. US Census Bureau 2010 Census TIGER/Line Shapefiles. http://www.census.gov/geo/www/tiger/tgrshp2010/tgrshp2010.html
17. Ernst, M., Lang, M., & Davis, S. (2011). Dangerous by design: Solving the epidemic of preventable pedestrian deaths, Transportation for America: Surface Transportation Policy Partnership, Washington, DC.
18. Ester, M., Kriegel, H. P., Sander, J., & Xu, X. (1996). A density-based algorithm for discovering clusters in large spatial databases with noise. *Kdd, 96*(34), 226–231.

19. MacQueen, J. (1967). Some methods for classification and analysis of multivariate observations. *Proceedings of the fifth Berkeley symposium on mathematical statistics and probability, 1*(14), 281–297.
20. Oliver, D., Shekhar, S., Kang, J. M., Laubscher, R., Carlan, V., & Bannur, A. (2014). A k-main routes approach to spatial network activity summarization. *IEEE Transactions on Knowledge and Data Engineering, 26*(6), 1464–1478.
21. Buchin, K., Cabello, S., Gudmundsson, J., Lffler, M., Luo, J., Rote, G., et al. (2010). Finding the most relevant fragments in networks. *Journal of Graph Algorithms and Applications, 14*(2), 307–336.

Chapter 4
Summary

Abstract This chapter summarizes the main challenges in finding a compact description or representation of observations on large spatial or spatiotemporal networks, i.e., the many connected components in the spatial network and the many candidates that have to be processed.

Many application domains may benefit from summarizing spatial network data including transportation safety, public safety, public health, and disaster response. For example, transportation planners and engineers may need to identify road segments that pose risks for pedestrians and require redesign, epidemiologists may try to understand the spread of disease so that patterns of progression can be established, and hydrologists may try to summarize significant environmental change on water resources to understand the behavior of river networks and lakes. However, the following computational challenges must be overcome when summarizing spatial network-based observations: (1) There may be a large number of k-subsets of connected components in the network (many connected components) and (2) There may be a large number of candidates (many candidates).

This brief reviewed a suite of techniques that address these computational challenges. First, spatial network activity summarization was explored that addressed the challenge of a large number of k-subsets of connected network components via the K-Main Routes (KMR) algorithm. Second, the challenge of the large number of candidates was examined by solving the significant route discovery problem using the Significant Route Miner.

Many Connected Components: The challenge of dealing with many connected components was conceptualized as the spatial network activity summarization (SNAS) problem. In SNAS, we are given a spatial network and a collection of activities (e.g., crime reports) and the goal is to find k shortest paths that summarize the activities. SNAS is important for applications where observations occur along linear paths such as roadways, train tracks, etc. SNAS is computationally challenging because of the large number of k subsets of shortest paths in a spatial network. The current state-of-the-art focuses on either geometry or subgraph-based approaches (e.g., only one path). An algorithm representative of a new trend in subgraph-based approaches, i.e., K-Main Routes (KMR), was presented. KMR discovers k shortest

© The Author(s) 2016
D. Oliver, *Spatial Network Data*, SpringerBriefs in Computer Science,
DOI 10.1007/978-3-319-39621-7_4

paths to summarize activities. KMR generalizes K-means for network space but uses shortest paths instead of ellipses to summarize activities. To improve performance, KMR uses network Voronoi, divide and conquer, and pruning strategies. A case study comparing KMR's network-based output (i.e., shortest paths) to geometry-based outputs (e.g., ellipses) on pedestrian fatality data was presented.

Many Candidates: The challenge of handling many candidates was conceptualized as the significant route discovery (SRD) problem. Given a spatial network and a collection of activities (e.g., pedestrian fatality reports), Significant Route Discovery (SRD) finds shortest paths in the spatial network where the concentration of activities is unusually high (i.e., statistically significant). SRD is important for societal applications in transportation safety, public safety, or public health such as finding paths with significant concentrations of accidents, crimes, or diseases. SRD is challenging because (1) there are a potentially large number of candidate paths ($\sim 10^{16}$) in a given dataset with hundreds of millions of activities or network nodes and (2) significance testing does not obey the monotonicity property. In a spatial network context, SaTScan may miss many significant paths since a large fraction of the area bounded by circles for activities on a path will be empty. Current network-based approaches only consider a small fraction of the network and only one significant network component (e.g., a path). Novel algorithms for discovering statistically significant linear hotspots using the ideas of likelihood pruning, Monte Carlo speedup, and dynamic segmentation were presented. The chapter also outlined a case study comparing various techniques on real data.

Printed in the United States
By Bookmasters